FROM PUGWASH TO PUTIN

FROM PUGWASH TO PUTIN

A Critical History of US-Soviet Scientific Cooperation

Gerson S. Sher

INDIANA UNIVERSITY PRESS

This book is a publication of

Indiana University Press
Office of Scholarly Publishing
Herman B Wells Library 350
1320 East 10th Street
Bloomington, Indiana 47405 USA

iupress.indiana.edu

© 2019 by Gerson S. Sher

All rights reserved

No part of this book may be reproduced or utilized in any form or by any means, electronic or mechanical, including photocopying and recording, or by any information storage and retrieval system, without permission in writing from the publisher.

The paper used in this publication meets the minimum requirements of the American National Standard for Information Sciences—Permanence of Paper for Printed Library Materials, ANSI Z39.48-1992.

Manufactured in the United States of America

Cataloging information is available from the Library of Congress.

978-0-253-04261-3 (cloth)
978-0-253-04262-0 (paperback)
978-0-253-04263-7 (ebook)

1 2 3 4 5   23 22 21 20 19

For Sasha

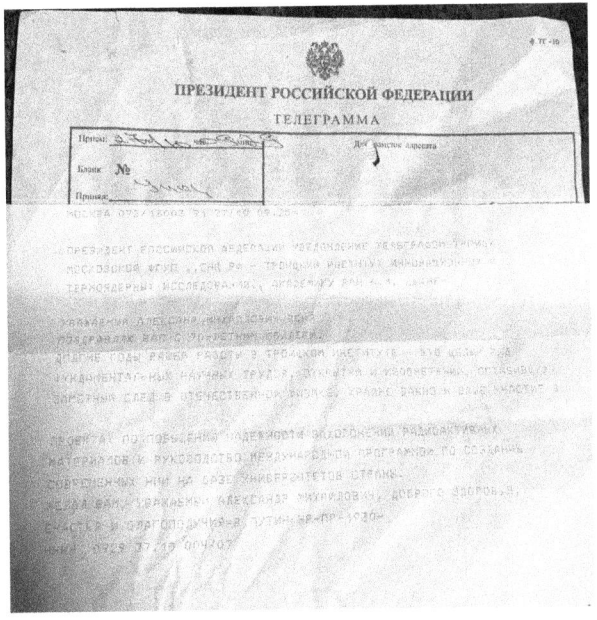

Translation

[Heading]
President of the Russian Federation
Telegram
[Date]

Dear Aleksandr Mikhailovich – I congratulate you on your 70th birthday. For many long years, your work at the Troitsk Institute have resulted in an entire series of fundamental scientific research, discoveries, and innovations, leaving a perceptible legacy in our land's physics. Also extremely important is your participation in projects on promoting the reliability of burial of radioactive materials and your leadership of an international program on creation of modern scientific research institutes on the basis of the country's universities. I wish you, Aleksandr Mikhailovich, good health, happiness, and well-being.

V. Putin.

This publication is based on work supported by a grant from the John D. and Catherine T. MacArthur Foundation. Any opinions, findings and conclusions, or recommendations expressed in this material are those of the author and do not necessarily reflect the views of the John D. and Catherine T. MacArthur Foundation.

CONTENTS

Acknowledgments xi

Introduction 1

Part I The Timeline

1 The Deep Cold War and the Exchange Programs 10

2 Détente and the Heyday of Massive Agreements 20

3 Sanctions and Perestroika 31

4 After the Fall: New Times, New Approaches 41

Part II In Their Own Words

5 How Did It Start? 75

6 What Kept Them Going? 113

7 Scientific Accomplishments 140

8 Other Accomplishments 165

9 Problems 187

10 On the Nature of Science in the Former Soviet Union 217

11 Vignettes 240

Part III Conclusion: So What?

12 What to Make of It All?   259

Appendix: List of Interviews   279

Bibliography   285

Index   295

ACKNOWLEDGMENTS

I AM NOT A SCIENTIST, CERTAINLY NOT TRAINED in the natural sciences, though some would oxymoronically call the discipline in which I earned my degree political science. My work has been the science education that I never had, and I've sought to take advantage of it. I thank countless scientists with whom I have been most fortunate to work—members of the National Academy of Sciences, the professional staff of the National Science Foundation, scientists throughout the Soviet Union and its successor states who have tolerated my visits in their laboratories and my questions, and American scientists and scientists of other nations who have in so many ways contributed to the management and oversight of the programs in which I've been involved.

In this regard, I wish to thank one very distinguished scientist in particular, though regrettably posthumously, Academician Aleksandr Mikhailovich Dykhne. Sasha, as he was known, was a giant among men and women of science.[1] In addition, his impeccable character was such that he was revered by all who knew him. This was demonstrated for me when our Russian colleagues insisted that only he could bring order to a potentially unruly binational expert panel for one of our hotly contested competitions. Beyond this, he was an exceptionally warm, modest, and caring person who, however, suffered no fools. I dedicate this book to his memory. One of the few regrets of my career is that I failed to convey a proper understanding to Sasha of the holy American sport of baseball, for which I take full responsibility—as well as for any errors of fact and interpretation that may appear throughout this book.

Professor Loren R. Graham, professor emeritus of the history of science in the Science, Technology, and Society Program at the Massachusetts Institute of Technology and associate of the Davis Center for Russian and Eurasian Studies at Harvard University, has been an inspiration, mentor, colleague, and friend for many years and has encouraged me and egged me on to write this book. Loren's 1972 book, *Science and Philosophy in the Soviet Union*,[2] played a major role in my intellectual development when I was studying the Yugoslav Praxis group of Marxist philosophers for my dissertation under another great thinker, the late political historian Robert C. Tucker, my primary dissertation advisor, and the indomitable Stephen F. Cohen, my second advisor.[3] Once I decided to make a profession of managing cooperative science programs, Loren's path and mine intersected frequently, most notably in our collaboration on the Basic Research and Higher Education program at CRDF (US Civilian Research and Development Foundation for the Independent States of the Former Soviet Union; now CRDF Global). It has always been a thrill to work as a peer, if that is the right term, with such a thoroughly distinguished scholar and to be able to bat around with him thoughts about Russia, science, and life. All the while, he also encouraged me strongly to pursue my scholarly interests. In the United States, we do not have schools of scientists or scholars in the Russian and German traditions, but if we did, there would be a Graham School of science history worthy of similar recognition.

The extensive number of interviews I carried out was possible only due to the generosity of the John D. and Catherine T. MacArthur Foundation, through a special travel and learning grant that was administered by CRDF Global. The MacArthur Foundation has been a key private supporter of scientific and scholarly cooperation with Russia, willing to undertake ambitious strategic initiatives—in particular, in research and education—that have borne fruit in many ways. While any opinions, findings and conclusions or recommendations expressed in this material are solely those of the author and do not necessarily reflect the views of the John D. and Catherine T. MacArthur Foundation, I wish to acknowledge here not only my personal gratitude but also my admiration for its visionary engagement in this broad field.

In addition, there are two friends without whom I would have been unable to coordinate my interviews in Georgia and Ukraine and who assisted me with follow-up correspondence: Viktor Los of the Institute of Magnetism of the National Academy of Sciences of Ukraine, whom I met many years before when he served as science attaché at the Ukrainian embassy in Washington, DC, and Helen Giorgadze of the Georgian Research and Development Foundation, which was created by CRDF Global and became a model for competitive grant making in that beautiful country. I am deeply grateful to them for their invaluable help.

I could not have written this book without the indulgence of the sixty-two remarkable people who allowed me to interview them. Their names are listed in the appendix. They were uniformly gracious, candid, and eager to share their experience in the hope that it might be of broader benefit. While the interviews themselves were generally no longer than an hour, the participants spent a great deal of additional time in both preparing for the discussion and in reviewing the sometimes voluminous excerpts that I sent them for approval. Their words, much more than mine, are what make this study unique, insightful, authentic, and at times even fun. Their testimony creates an important historical record of a massive, historic undertaking. I hope that they find this book to be an adequate vehicle for others to understand and reflect on their rich experience and assessments. Some of the quoted passages from the interviews are lengthy, but I have left as much text as possible intact because of the richness of personal detail that gives the speakers' words special meaning.

I am also grateful to those who have made comments or suggestions on portions of my manuscript: Murray Feshbach, Loren Graham, Eric Green, Maija Kukla, John Malin, Norman Neureiter, the late Arthur E. Pardee Jr., Peter Reddaway, and Valery Soyfer. My acquisitions editor at Indiana University Press, Jennika Baines, is responsible for urging me to put the text into readable form.

Words do not suffice to express my deepest debt of gratitude, to my family. More than any person on earth, my wife, Margery Leveen Sher, the love of my life, has been my best guide, critic, editor, and friend in all things. About life, I have learned some of my most valuable lessons from my children, Rabbi Jeremy D. Sher and Adam Leveen Sher, who have taught me by their deeds to cherish the unexpected in ourselves and in others. The list of other friends and colleagues from whose counsel I have benefited is simply far too long to enumerate, but you know who you are.

All that having been said, I take full responsibility for any mistakes, misunderstandings, misrepresentations, and the like that may appear in this work.

Notes

1. See the impressive appreciation by his many distinguished colleagues in Alfimov, et al. 2005.
2. Graham 1972.
3. The dissertation was later published as Sher 1977.

FROM PUGWASH TO PUTIN

INTRODUCTION

The story of scientific cooperation between the United States and the former Soviet Union (FSU) is filled with science but also with human drama. It is a story of prominent scientists from both countries who spoke a common language—the language of science—and who saw a compelling opportunity not only to advance knowledge but also to achieve global peace in a world made terrifying, partly through their own doing. It is a story of governments seeking to create common interests through people-to-people contacts while seeking advantage in the realms of international affairs, security, and economic well-being. It is a story of how these motivations strengthened and empowered each other but also of how they occasionally came into conflict. It is a story of the blush of idealism as well as the disappointments of ideals shaken by harsh encounters with national interest, international conflict, and public (sometimes willful) misunderstanding of the goals and methods of the cooperation.

Like any good story, it is rich in motifs and subplots. One of the main themes we will encounter is the contest between the open and closed society. These terms are less a completely accurate description of reality, for no society is ever completely open or completely closed, than they are helpful characterizations.[1] One of the most popular ideas about scientific cooperation between the two superpowers was the notion that in the contest for scientific, technological, and military advantage, the closed society wins. But is this really true? Was the United States really taken to the cleaners by the Soviet Union through its cooperative science programs, providing invaluable opportunities for our adversaries to steal our secrets while keeping under wraps its own secret institutes? Did we really surrender our most precious technological treasures to the Soviets for the sake of some fuzzy vision of global harmony or in exchange for some throwaway concessions meant to hold the Soviets accountable in other realms? It will be up to the reader, based on the direct, firsthand testimony offered in this book, to decide whether this narrative needs to be reexamined.

Another motif is the notion that "science knows no boundaries." This is a refrain that we often hear from scientists themselves with regard to their own work and role as actors on the international stage. But is it really true? Does it follow that because modern science is in agreement on methods and standards of experimentation and proof—that in this sense there is a universal language of science—scientists are entitled to consider themselves exempt in some larger sense from the constraints of ordinary mortals? Or that science itself is at its best when there are no boundaries, no borders, no systematic impediments to perfect, unrestrained, frictionless communication? While in important ways the story of US-FSU science cooperation offers testimony to the passionate commitment of scientists to unfettered and transparent communication, it also suggests that the very boundaries that divide them—political, cultural, even perhaps linguistic—often enrich science with fresh insights and approaches.

This is also a story of a grand experiment in the world of international scientific cooperation itself. As an historical phenomenon, bilateral, government-driven scientific cooperation between the two post–World War II superpowers was a new and untested concept. It represented the very first time that the concept of purposeful bilateral scientific cooperation between two major countries became enshrined as public policy doctrine and implemented in practice on a very large scale. For centuries, scientists had collaborated in the pursuit of knowledge, often traveling extensively and freely to work in each other's laboratories and to share insights and accomplishments at scientific meetings. In many cases they became close friends. This, however, was the first time that formal bilateral programs of science cooperation—created by governments as matters of foreign policy—came into being. In this important sense, it was the antecedent of all modern bilateral scientific cooperation. This new form of scientific cooperation was artificial from the outset, though it was virtually the only way that scientists from the two superpower adversaries could work together. Even in quantitative terms—the number of scientists involved, the immense scale of resources invested, the significance and breadth of issues addressed, and the political visibility of these efforts in both countries—it was unprecedented and is unlikely to be repeated.

This was an experiment in many dimensions and in reality a series of experiments, each corresponding to its own time in the course of the twentieth century. Of this experiment, we do not yet know the final outcome or even what standards we should use to gauge the outcome. But enough time has passed—sixty years since the experiment's launch at the height of the

Cold War—to be able to look back and say something about what we have learned from it.

There are many excellent scholarly studies of the history of US-FSU scientific cooperation.[2] My goal is somewhat different. It is to tell the story of this remarkable effort, and I do this primarily through the stories told to me by sixty-two people, from both the United States and the FSU, who participated in the programs as scientists, designed and managed them in government and nonprofit organizations, and sought as diplomats to relate them to national foreign policy goals.[3] At times I have inserted stories from my own forty-year experience as a manager and leader of programs in the government and nonprofit fields. I was extraordinarily fortunate to be able to interview individuals whose experience spanned the entire six decades of US-FSU science cooperation, reaching back as far as the first scientists to make long-term exchange visits in the Soviet Union in 1959. In this, it is also an important historical record. It is these stories within stories that make this book unique as a kind of oral history and that have made it a pleasure to research and write.

This is a book about science, but it is not a scientific book. It's about international relations, but it does not delve into theories of international relations, though it harshly criticizes one of its most popular avatars. It is a history of an important and unique scientific endeavor, but it is not a classical history of science. It draws on all these fields, and I hope it will be of interest to specialists from all these fields; however, it is also very much intended for the general reader who is interested in a good, real-life story. It draws, most importantly, on the personal memories and observations of distinguished individuals from all these fields and how their experiences helped to inform their own work. While there are some discussions of technical issues that I have felt necessary to include in order to illuminate the importance of certain outcomes—in particular, scientific findings—the general reader may gloss over these while still getting a sense of why they were so important.[4]

The first part of the book, chapters 1 through 4, traces the history of the cooperative science programs between the United States and the FSU in the post–World War II period. It is preceded by a timeline for convenient reference. Rather than give a dry recounting of events, I have also sought to outline the main ideas and goals of these programs as they evolved and have included the personal recollections of many of the actual actors whom I was privileged to interview. Part II, chapters 5 through 11, is where we delve into the more personal world of our witnesses—how they got involved and why, what the problems they encountered were, and what their accomplishments were, as well as what we have learned from them about how science was done in the FSU. In Part III, I attempt to answer the big question—so what?—and to offer my thoughts about lessons learned.

On with the story.

Gerson S. Sher
Washington, DC
May 21, 2018

Notes

1. In the terms of sociologist Max Weber, they are "ideal types" of not only the institutions but also the ideas and cultures around which societies are organized.

2. As the study of national science systems goes, Russian and Soviet science must be one of the most extensively studied in the world. In this note, I am citing only some of the most recent studies. Balzer and Sternheimer 1989; Graham 2016; Graham 2013; Graham and Dezhina 2008; Schweitzer 2013; and Soyfer 2002.

3. For my interviews in the former Soviet Union, I traveled to Ukraine and Georgia for a total of nearly three weeks meeting scientists who had active collaborations with US colleagues in both the Soviet and post-Soviet eras, as well as former government officials with whom I had worked in the latter period. I did not travel to Russia, much as it would have unquestionably enhanced my research to do so. However, early in the process of writing to potential interlocutors in Russia—people I had known well and worked with closely as colleagues for many years—I detected a disturbing pattern. Either they did not respond, uncharacteristically, or in one particularly surprising case, the individual at first agreed to at least a video interview and then expressed concern through a mutual friend that my questions might be too sensitive. This struck both me and my friend as strange, since the questions themselves—"How did you get involved in bilateral science cooperation with the United States, and why?"; "How did it go?"; "How do you retrospectively evaluate your experience in light of your initial expectations?"; and "What are your favorite stories?"—were very neutral. Subsequently I wrote reassuringly to that individual and attempted to schedule a discussion. He never wrote back, also very uncharacteristically. After sixty-four visits to Russia spread over forty-five years, it was very clear to me that the atmosphere in that country had become so hostile to candid discussion of this topic that to travel there would be not only pointless but also potentially risky and counterproductive. To compensate somewhat, I did seek out émigré Russian scientists living in the United States and interviewed several; moreover, to some extent the experience of scientists from Ukraine and Georgia helped to fill in some of the voids. While I regret that I did not travel to Russia for the book, I regret even more that the veil of repression and paranoia has returned to that country with such force as to make the trip unadvisable.

Of the sixty-two interviewees, thirty-seven were scientists; fourteen, government officials; six, nonprofit managers; three, from the commercial sphere; and two, scholars. Thirty-five were from the United States and twenty-seven from the former Soviet Union, including eleven from Ukraine, ten from Georgia, and six from Russia, whether they resided in Russia or in the United States (the latter were émigrés who had worked for substantial periods of time in Russia).

4. Some brief notes about usage: I use the term *former Soviet Union*, or FSU, to refer to both the Union of Soviet Socialist Republics and collectively the countries that arose from the ashes of that empire after 1991. In a practice that may at first seem confusing to the reader, I sometimes use the words *Russian* (or *Ukrainian*) and *Soviet* in a way that may appear interchangeable; in fact, I deliberately use these words to refer not to geography but to historical context. For example, when talking about the role of secrecy in science, I will use the word *Soviet* because the condition of secrecy pertains in particular to the Soviet period and its legacy; on the other hand, when talking about a *Russian* tradition in science, for example, the tradition would be one that predated the Soviet period, with its roots in the history of the Russian Empire. In transliterating Russian-language names and other words, I use as a default the standard scientific transliteration system, in which, for example, the name Ивановский is rendered as "Ivanovskiy," with the ending, -*skiy*. For surnames that have come into common English usage, such as Достоевский, I use the standard English usage—that is, "Dostoevsky," without the intervening *i*. In referring to the United States of America, I use the shorthand *US* as an adjective to denote both government and private entities, such as "US scientists," which could include both government and nongovernment scientists, though sometimes I also use the term *American*, by which I refer to the United States of America, not necessarily to all of the Western Hemisphere; when discussing US government entities, I will identify them specifically as such unless in a direct quotation.

PART I
THE TIMELINE

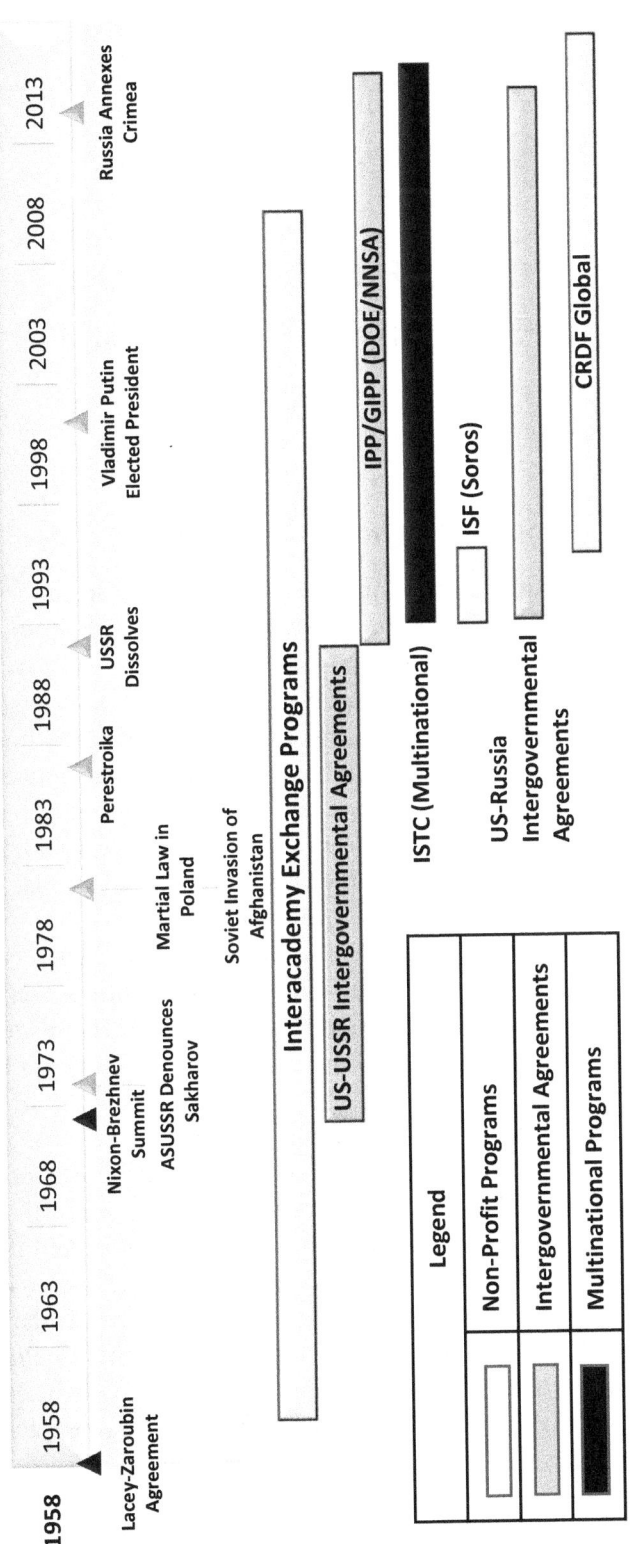

Figure 1.1. The Timeline

1

THE DEEP COLD WAR AND THE EXCHANGE PROGRAMS

The Spirit of Pugwash, the Thaw, and *Sputnik*

The story of scientific cooperation between the United States and the former Soviet Union begins not with a program but rather with an understanding: that scientists from the two postwar superpowers who developed the most destructive weapons in human history had a responsibility to see that these very weapons were not used to destroy humanity. This understanding became known as the "Spirit of Pugwash," named for a village in Nova Scotia, Canada, where magnate Cyrus Eaton kept his summer home. There, twenty-two prominent scientists from the United States, the Soviet Union, and eight other countries (Japan, the United Kingdom, Canada, Australia, Austria, China, France, and Poland) gathered in 1957 to discuss the threat to humanity posed by the advent of nuclear weapons.[1] Such a conference had been urged in 1955 in what came to be known as the Russell-Einstein Manifesto, eleven of whose twelve signatories were Nobel Laureates, including Bertrand Russell and Albert Einstein, who signed just days before his death.

The Russell-Einstein Manifesto and the Spirit of Pugwash, however, were not the first efforts to address these grave concerns. Scientists such as Einstein and Robert Oppenheimer were obsessed with worry that the creation of atomic weapons would present a destructive potential that would be a profound threat to humanity itself. In 1945, Eugene Rabinowitch of the University of Illinois, who participated in the work of the Manhattan Project, along with Hyman Goldsmith, founded the influential *Bulletin of the Atomic Scientists* with the mission of educating the public about the dangers of nuclear war. Rabinowitch was also a key figure in Pugwash, becoming a founding member of the Continuing Committee for the Pugwash Conferences on Science and World Affairs.[2]

Underlying the Pugwash movement was a sense that scientists (and other intellectuals) have a special responsibility to direct their activity to peaceful purposes. Well aware of their role in creating new weapons that could destroy humanity, they saw it as their mission to establish direct lines of communication with high-level policymakers and to urge them to harness scientific knowledge in beneficial, not destructive, ways. The Spirit of Pugwash in the late 1950s had broad appeal to leading scientists in the United States and the Soviet Union. It created a shared view that peaceful coexistence and global security depended significantly on dialogue and cooperation between these two superpowers. It also converged to a certain extent with some US policy initiatives, such as the Atoms for Peace Program, and discussions between diplomats of the two countries to create citizen-to-citizen cultural and scientific "exchange programs," which were established in the late 1950s.

Within the Soviet Union, things were also changing in the early 1950s, though at first less perceptibly. Joseph Stalin's death on March 5, 1953, marked the end of an era of terror and a monstrous "cult of personality," as Nikita Khrushchev was soon to call it.[3] But to all except the most avid Kremlinologists, the waters seemed calm and the monolithic rule of the Communist Party of the Soviet Union (CPSU) appeared intact, even in the vacuum left by the death of the great *Vozhd'* ("the leader," one of Stalin's many heroic epithets).

This magic spell ended in February 1956 with the Twentieth Congress of the CPSU. At that time, Khrushchev made his historic, dramatic denunciation of Joseph Stalin before a stunned assemblage of the party's leaders, who had only three years before gone into deep mourning with the entire nation over Stalin's death. Khrushchev's speech, in which he also debuted the notion of what came to be known as "peaceful coexistence,"[4] launched a far-reaching thaw in Soviet society, setting the stage for reconciliation and conflict on a number of fronts—the return of prisoners from the gulag and some cultural relaxation but also conflict with China, an attempted reactionary coup in 1957, and ultimately Khrushchev's own demise in 1964.

Khrushchev's 1956 speech was also a signal to the West that some relaxation in international relations might be possible. Consequently, it was now possible to talk seriously about initiating formal cultural relations between the United States and the USSR.[5] Over that summer, President Dwight

D. Eisenhower met with a group of citizens from different fields to discuss the possibility of direct person-to-person exchanges between the two countries. The process culminated in a September 1956 conference in which Eisenhower formally announced the formation of the People-to-People Program. Explaining his vision for the program, he said, "If we are going to take advantage of the assumption that all people want peace, then the problem is for people to get together and to leap governments—if necessary to evade governments—to work out not one method but thousands of methods by which people can gradually learn a little bit more of each other."[6]

This philosophy formulated the primary rationale for the exchange programs of the ensuing period: people-to-people contacts. Another term that appeared early to describe the same concept, and one that has stuck, was *citizen diplomacy*. According to the Institute of International Education, an organization that has been the country's leader in managing person-to-person exchanges for almost one hundred years, "Citizen diplomacy is a concept that involves two seemingly disparate ideas: private citizens engaging in individual endeavors that serve their own interests; and diplomacy, which includes a framework for cooperation between countries. Taken together, citizen diplomacy refers to an array of actions and activities that individuals can partake in that contribute to deepening ties between individuals and communities and to advancing the goals of public diplomacy. Citizen diplomacy is thus an integral part of public diplomacy."[7]

One of the most popular books of the late 1950s, *The Ugly American*, whether deliberately or not, brought the idea of citizen diplomacy to the public's attention. It drew a stark contrast between the lives and work abroad (in this case, in Southeast Asia) of American diplomats, whom it depicted as oblivious to and contemptuous of local conditions, and the few "ordinary" American citizens, who came to these countries to work on the ground in villages, to listen to the residents' ideas and needs, and to learn the language and culture. While the eponymous "ugly American" was the latter kind of person—and as portrayed in the book, literally a physically unattractive man—it was the polished diplomatic corps of the local US embassy who were truly "ugly" in terms of their behavior and disrespect for local people and culture. One of the book's characters, the fictional Philippines minister of defense, observed to the American ambassador of a fictional nearby country: "The simple fact is, Mr. Ambassador, that average Americans, in their natural state, if you will excuse the phrase, are the best ambassadors a country can have. . . . They are not suspicious, they are eager to share their skills, they are generous. But something happens to most Americans when they go abroad. Many of them are not average.... They are second-raters."[8]

One can only imagine the ambassador's reaction to the minister's characterization of American diplomats and vacation tourists as "second-raters." But these words faithfully underscore what Eisenhower meant by "leaping governments" with ordinary people as citizen-ambassadors. And it was his faith in the goodness of the ordinary American that informed his ideas on exchanges.

The launch of *Sputnik* on October 4, 1957, was a shock and a wake-up call to the West about the Soviet Union's apparent scientific and technological prowess. Unlike the Soviet development of the atomic and hydrogen bombs, in which it was thought (and possibly was true) that the Soviets merely copied designs from the West through espionage, no similar case could be made for the sudden appearance of the first artificial satellite in human history orbiting around the Earth. It was clear from *Sputnik* that the Soviets were a power to be respected and feared in terms of their scientific and, in particular, engineering capabilities after all. *Sputnik*'s launch was in effect the beginning of the two races that characterized the second half of the twentieth century—the space race and the missile race.[9] American scientists were now challenged to fill in vast gaps in their knowledge about what was going on in Soviet research institutes, and the study of Russian science—as well as the Russian language, for that matter—became a national security issue.

The Lacy-Zaroubin Agreement

On January 27, 1958, the United States and the Soviet Union signed the "Agreement between the United States of America and the Union of Soviet Socialist Republics in the Cultural, Technical, and Educational Fields." The agreement was signed by the two chief negotiators—William S. B. Lacy, President Eisenhower's special assistant on East-West exchanges, and Georgi Z. Zaroubin, the Soviet ambassador to the United States—and became known as the Lacy-Zaroubin agreement.[10]

The Lacy-Zaroubin agreement (often shortened to simply "Lacy-Zaroubin" or "the Cultural Exchanges agreement") was a major innovation in international cooperation of any kind. As Yale Richmond, who oversaw the US-Soviet programs for decades in the State Department's Bureau of Cultural Affairs, emphasizes in his history of the early exchange programs, "on the U.S. side there was no precedent for such an agreement."[11] After the

Second World War, he explains, the United States had unilaterally funded and administered exchange programs with Germany and Japan with the goal of democratizing these countries, but the notion of a formal bilateral, intergovernmental program was new.[12] Also new was the involvement of the US nonprofit and private sectors in the program, which encompassed science and technology, broadcasting, motion pictures, youth, education, performing arts, athletics, and tourism. In some cases, the activities were almost completely self-funded and self-administered. "Why, then," Richmond asked, "was such an agreement necessary?" "The simple answer," he continued, "is that the Soviet leaders wanted an agreement and made it a condition to having exchanges."[13] In short, it was to satisfy the Soviets' obsession with control. All the areas covered by the exchanges, in the Soviet Union, were state funded and state controlled.

Activities taking place under this framework were highly structured. The very term *exchange* brought to mind a carefully calibrated and planned transaction, not unlike an exchange of spies, as a matter of fact. Typically, they functioned through quantitative quotas, expressed in terms of persons or person-months.[14] The programs' basic underlying principles, as expressed in the agreement, were equality, reciprocity, and mutual benefit.

The US and Soviet governments' policy objectives were in some respects similar and in others quite different. In his sweeping study of this period, Richmond describes them as follows:

> U.S. objectives, as stated in a National Security Council directive (NSC 5607), were, among others, to broaden and deepen relations with the Soviet Union by expanding contacts between people and institutions; involve the Soviets in joint activities and develop habits of cooperation with the United States; end Soviet isolation and inward orientation by giving the Soviet Union a broader view of the world and of itself; improve U.S. understanding of the Soviet Union through access to its institutions and people; and obtain the benefits of long-range cooperation in culture, education, and science and technology.
>
> The Soviet objectives in the exchanges were not openly stated but, from a study of how they conducted the exchanges, they can be presumed to have included the following: to obtain access to U.S. science and technology, and learn more about the United States, its main adversary; support the view that the Soviet Union was the equal of the United States by engaging Americans in bilateral activities; promote the view that the Soviet Union was a peaceful power seeking cooperation with the United States; demonstrate the achievements of the Soviet people; give vent to the pent-up demand of Soviet scholars, scientists, performing artists, and intellectuals for foreign travel and contacts; and earn foreign currency through performances abroad of Soviet artists.[15]

Lacy-Zaroubin—renegotiated, revised, and modified over the years—remained the diplomatic and legal underpinning for all major US-Soviet cultural and scientific exchange and cooperation activities until the Soviet Union's dissolution in 1991. It provided the framework for an enormous range of people-to-people activities, and this formal framework agreement was cited in every significant subsequent intergovernmental agreement in this area as a matter of protocol. These included not only science and technology and scholarly research in the social sciences and humanities but also myriad other activities, including celebrated artistic performances, traveling exhibits (including the American exhibit that formed the venue for the famous Nixon-Khrushchev "kitchen debate" in 1959), and exchanges of professionals and ordinary citizens. All these activities took place within the carefully defined parameters of a bilateral agreement that was as politically sensitive as it was unprecedented.

The Exchange Programs

Even before the 1958 signing of Lacy-Zaroubin, the American scholarly community saw an opportunity to begin professional contacts with the thaw in the Soviet Union set in motion by Khrushchev's dramatic February 1956 denunciation of Stalin. In that very month, a group of scholars formed the Inter-University Committee on Travel Grants, or IUCTG, to promote the exchange of scholars between the United States and the Soviet Union.[16] Funded by the Ford Foundation, the program enabled visits of up to one month in either country, but only with "tourist" visas. Lacy-Zaroubin made it possible for these and other visits to take place under a more formal and appropriate framework, with special visas for exchange visitors and introducing the policy of "receiving-side pays," which also greatly simplified financial arrangements.

Once Lacy-Zaroubin entered into force, the State Department asked IUCTG to conduct the official exchange program. (In 1968, IUCTG transferred these programs to the newly founded International Research and Exchanges Board, or IREX.) The IUCTG/IREX program was the most durable and longest-lasting exchange or cooperative program of any kind, surviving even the breakup of the Soviet Union, but it eventually terminated its presence in Russia in 2015 under relentless pressure and harassment from the government of Vladimir Putin. With the advent of such programs specifically oriented toward science and technology over the following years, IREX shifted its emphasis increasingly to the humanities and social sciences and expanded its scope to many other countries. Its remarkable history is thus beyond the purview of this book.[17]

The addition of the natural sciences to the array of exchange programs came later, in 1958, after the shock of Sputnik. It was against this background that the Lacy-Zaroubin also called for a program of scientific exchange visits to be organized by the US National Academy of Sciences (NAS) and the Academy of Sciences of the USSR (ASUSSR).[18] NAS president Detlev Bronk, who had participated in the Lacy-Zaroubin discussions in 1958, met again with Soviet Academy of Sciences representatives in early 1959 and, in November of that year together with Soviet Academy of Sciences president A. N. Nesmeyanov, signed the first detailed interacademy scientific exchange agreement.[19] By 1977, some 350 American and an approximately equal number of Soviet scientists had participated in the program, a pace that continued for several more years.[20] Also, beginning in the mid-1960s, this model was extended to exchanges with the academies of sciences of the countries of eastern Europe. In the years prior to the era of détente inaugurated by the Nixon-Brezhnev summit meeting of 1972, the interacademy program was the primary formal mechanism allowing scientists from both countries to visit and, in some cases, collaborate with each other professionally.

The interacademy program was the longest-lasting scientific exchange program with the Soviet Union and later Russia, lasting thirty years. Funded by the National Science Foundation, the interacademy program began losing support in the 1980s—as relations between the United States under President Ronald Reagan and the "evil empire" had reached an all-time low—and the programs were finally phased out in 2009. During this period, however, the interacademy exchanges continued to play an extremely important, if reduced, role as the sole national, structured bilateral instrument available to scientists in both countries to pursue their scientific interests on an individual basis.

While the exchange programs were of enormous benefit in terms of opening up serious professional contacts between scientists and scholars of the two countries, they were not without problems. One of the main issues that was much remarked on at the time was the programs' formality, which was at variance with scientists' previous experience of free, unfettered movement throughout the world. The idea of formal bilateral scientific cooperation between two specific countries, governed by carefully negotiated diplomatic agreements, was not part of that tradition. The Lacy-Zaroubin provisions governing cultural and scientific US-Soviet exchanges were fundamentally different from these norms. They were rigid, structured, highly specific, and restrictive. They prescribed everything, from the areas of cooperation to the number and duration of visits, their financial and legal frameworks, and more.

It was a very unusual situation for American scholars and scientists, who were used to making their own arrangements as a traditional right and matter of convenience. In contrast, Lacy-Zaroubin had explicit quotas, which the Americans also found very frustrating and distasteful. Professor Robert Byrnes of Indiana University, the founding cochair of the IUCTG, once complained that the two countries were exchanging students "like so many sacks of grain."[21]

Turning to the interacademy program, the implementing agreement between the two academies (under the general umbrella of Lacy-Zaroubin) that governed it was so highly structured that its appendixes listed in careful detail individual "desirable" visits by Soviet and American scientists in the opposite country by subdiscipline. The Americans, as usual, found all this formality extremely bothersome and unnatural. Indeed, even the sponsoring institutions, the NAS and the Soviet Academy, were quite different. Both were honorific institutions, meaning that they maintained a membership of highly distinguished scientists who were elected directly by the membership. This feature, in fact, sharply distinguished the Soviet Academy from virtually any other Soviet institution and was that body's principal claim to being a unique, autonomous institution in that country.

Beyond this, their differences were very pronounced. The NAS was essentially a distinguished scientific body created under President Abraham Lincoln to advise the government on matters of scientific and technical import. The Soviet Academy, on the other hand, had been established under Peter the Great (as the Russian Academy of Sciences) as a research institution and further transformed by Joseph Stalin into the USSR's primary scientific research institution with hundreds of institutes, in part because he distrusted the universities and in fact downgraded them to pedagogical institutions.[22] The NAS, a nonprofit institution, aside from its research panels and reports, conducted virtually no operational activities, and its finances, while largely from the government for specific studies, were independent; its main role was to provide independent advice to the US government. The ASUSSR, in contrast, was a massive operational system, theoretically independent but funded directly by the government, largely for military research. After the fall of the Soviet Union in 1991, it also came to be widely criticized as the largest property-owner in the Russian Federation.

These incongruities between the two academies were reflected in the discomfort that the Americans found with the exchange program arrangement. Lawrence Mitchell, the interacademy program's first NAS manager, acknowledged in the February 1962 *Bulletin of the Atomic Scientists* that the program's framework was "unusual" for an institution like the NAS. The NAS, Mitchell said, nevertheless agreed to it "in the hope that increased contacts . . . would hasten the day when scientists of both countries could visit each other at will, as is customary in the international scientific community."[23]

A major NAS review of the interacademy exchange program chaired by NAS member and MIT professor Carl Kaysen (the Kaysen Panel Report) in 1977 observed that "the institutionalization of the inter-academy exchange program posed serious problems from the outset and these tended to undercut the American side's goals and methods. Not only were the two academies quite asymmetrical as institutions; this was a strange way to go about exchanging scientists. A system of agreeing to fields for visits, length of research sojourns without regard to who would make the visit, who would receive the visitor, and the like was totally alien to the methods by which scientists of the industrially developed countries associate."[24]

These are points well taken. The formality and rigidity of the exchange programs, and their cumbersomeness and formulaic quality, were distasteful to scientists who were accustomed to traveling and associating freely with their colleagues abroad. Yet under the circumstances, they had very few options for visiting the Soviet Union or having their Soviet colleagues visit them. One of the key issues was visa issuance. The rigid system of quotas and bureaucratic oversight, codified by the governments to address concerns about security and symmetry of benefits, provided them with some level of assurance of control. These were all real and legitimate concerns that simply did not exist with regard to scientists of "the industrially developed countries."

As a program officer in the interacademy program in the 1970s, I used to joke that I was a glorified travel agent, responsible for drawing up itineraries for Soviet scientists and getting them cleared with the State Department; writing letters and negotiating with hosts on behalf of foreign exchange visitors; helping visitors from the USSR and eastern Europe understand the mysterious concept of a travelers' check; and getting American scientists ready to go overseas and deal with the exigencies of life in a strange country. But I also liked to say that if it were so easy for scientists to travel between the two countries, they wouldn't need people like me to do it. The same was no doubt true on the larger scale: If the two countries involved were any but the USSR and the United States of America in the depths of a cold war, they wouldn't need cumbersome formal agreements to enable the free exchange and movement that scientists viewed as their just right and privilege.

Notes

1. "About Pugwash" 2015.
2. Grodzins and Rabinowitch 1963.
3. Khrushchev 1956.
4. See Khrushchev 1959 for his comments on the concept of peaceful coexistence. It is generally agreed that he first used the term in his address to the 20th Congress of the CPSU.
5. In general on this period, see Richmond 2003, Richmond 2013, and Byrnes 1976.
6. D. Eisenhower 1956.
7. Bhandari and Belyavina 2011, 3.
8. Lederer and Burdick 1958, 108.
9. For an exhaustive study of *Sputnik*'s impact, see Brzezinski 2007.
10. For the full text, see "Text of Lacy–Zaroubin Agreement, January 27, 1958."
11. Richmond 2003, 16.
12. There is ample evidence, however, that the Soviets had a strong preference for formal exchange programs dating back to the 1930s, as Michael David-Fox (2012) has written in his study of Western visitors to the Soviet Union in the interwar period. For one thing, the more formal and structured the program, the easier it was to control, and since the purpose of these 1930s programs described by David-Fox was to showcase the achievements of Soviet "Socialism in One Country" for Western sympathizers, careful control was essential. This penchant for control certainly was a factor later on in the Soviet preference for highly structured, formal exchange programs in the humanities and the sciences.
13. Richmond 2003, 16.
14. A "person-month" refers to the amount of time an individual person spends in a given activity for a period of one month.
15. Richmond 2003, 17.
16. The following account is based largely on Richmond 2003, 22–23.
17. On a personal note, I owe my entire career in an important sense to Professor Allen H. Kassof of Princeton University, who, as the longtime director of IREX and my graduate sociology professor, suggested I look at the NAS's exchange program for job opportunities.
18. "Text of Lacy–Zaroubin Agreement," section 9.
19. The text of the 1959 interacademy agreement can be found in Schweitzer 2004, 104–12.
20. Lubrano 1985, 54.
21. Richmond 2003, 24.
22. For this story, see Graham 1977.
23. Mitchell 1962, 17.
24. *Review of U.S.-USSR Interacademy Exchanges and Relations*, 41.

2

DÉTENTE AND THE HEYDAY OF MASSIVE AGREEMENTS

SHORTLY AFTER PRESIDENT RICHARD M. NIXON'S FEBRUARY 1972 historic summit meeting in Beijing, on May 22–30 of that year he held an equally historic summit meeting in Moscow with general secretary of the Communist Party of the Soviet Union (CPSU) Leonid M. Brezhnev. With this meeting, Nixon and Brezhnev gave birth to the era of détente. It was at this time that the most significant expansion of science and technology cooperation between the two countries was born.[1]

The philosophy behind this new era of scientific cooperation was quite different from the one that preceded it. President Eisenhower, as we have seen, had stressed the importance of people-to-people exchanges to enable individual citizens to "leap" and even to "evade" governments in learning more about each other. In the more instrumental thinking of President Nixon, however, with a strong assist from Secretary of State Henry Kissinger, the idea was essentially to use science and technology cooperation as a direct instrument of foreign policy. Testifying before the Senate Foreign Relations Committee in late 1972, Kissinger acknowledged that the primary rationale for these agreements was political, not scientific. Before 1972, he claimed, "there were no cooperative efforts in science and technology. Cultural exchange was modest. As a result, there was no tangible inducement toward cooperation and no penalty for aggressive behavior. Today, by joining our efforts even in such seemingly apolitical fields as medical research or environmental protection, we and the Soviets can benefit not only two peoples but all mankind. In addition, we generate incentives for restraint."[2]

In this final sentence, we can see the source of an ambiguity that was to persist over the next two decades in the relationship between science and foreign policy. Science and foreign policy now became intimately related, not only in the sense of science and scientists helping diplomats to find solutions to problems of a transnational or international character, but also of science as a blunt tool of foreign policy, to be offered as an incentive for good behavior and withdrawn as a punishment for misconduct, in the expectation that such carrots and sticks would actually make a difference. Moreover, to the extent that the US-USSR intergovernmental science and technology agreements were models for the many similar agreements that followed, the ambiguity and potential confusion surrounding their purpose became a sort of contagion that permeated the public's understanding—including that of the scientific community—of the proper relationship between science and diplomacy. I will return to this central point toward the end of this book.

Prominent among the concrete instruments that Kissinger and Nixon designed to promote Soviet restraint were a slew of major agreements. The most notable and memorable of these was the Strategic Arms Limitation Treaty (SALT), signed at the Nixon-Brezhnev summit meeting in Moscow on May 26, 1972. Science and technology cooperation, however, was also a major topic of discussion and action at this meeting.

Norman Neureiter was a policy adviser at the Office of Science and Technology (OST),[3] informally called the White House Science Office, in the Executive Office of the President at the time. A decade earlier, Neureiter recalled, President John F. Kennedy, at a White House dinner for the prime minister of Japan,

> raised his glass in a toast and proposed the creation of three committees—an economic committee at the cabinet level, an academic committee drawn from universities, and then a third committee: for the first time ever, a president of the United States proposed a joint science committee, using science as an instrument of constructive international engagement. It was the president's answer to Ambassador Reischauer's challenging essay about "the broken dialogue" between the intellectual communities of the US and Japan. The instrument was to be scientific cooperation between US and Japanese scientists. The structure and details or the program were worked out by the Joint Committee over a number of months.[4]

Shortly thereafter, Neureiter was hired by the National Science Foundation (NSF) to manage the resulting intergovernmental bilateral cooperative science program with Japan, The experience made a deep impression on him. According to his account, the agreement with Japan proved to be a precedent for an overture a decade later to another country, the Soviet Union, with which the United States wished to repair relations. In 1971, Neureiter related,

> Unknown to me and I believe unknown to my boss, President Nixon's Science Advisor and Office of Science and Technology Director Edward David, Henry Kissinger had been soliciting ideas from several USG [US government]

agencies for science cooperation activities with the Soviet Union. The idea was to find ways in which the US and USSR could cooperate in some peaceful scientific activities. Also unknown to us then was that the two countries were planning a summit meeting sometime in 1972. Coincidentally, about a year earlier, I had proposed the creation of a Joint Committee on Scientific Cooperation between the US and the Soviet Union—something like the US-Japanese model I mentioned earlier. I thought there should be some examples of peaceful scientific cooperation between the US and USSR. It took almost a year to get all the necessary approvals for such an idea, but finally it was OK'd and included in what turned out to be the package of proposals presented to and agreed on by the Soviets at the Nixon-Brezhnev summit meeting in Moscow in May 1972.[5]

Neureiter's proposal for a science and technology agreement with the Soviet Union was also stimulated in part by Kissinger's visit to China in 1971. Prior to that trip, according to Neureiter,

> In a meeting in early 1972 with Henry Kissinger, President Nixon's national security advisor, my boss, OST Director and Science Advisor Edward David, was told that the president was planning a secret trip to China. In addition to the diplomatic discussions, he wanted to offer the Chinese something more than the geopolitical repositioning that would occur—something more tangible, more concrete; perhaps, he said, some proposals for science cooperation [that will] show the Chinese we are serious about some kind of enduring engagement. And he wanted very soon a package of science initiatives that could be taken to China for the discussions. Ed David told me to get to work on this right away; the package was needed in just a few days, and of course, no one was to be told about this plan. It had to be kept a secret.
> There was a committee that had been set up in the National Academy of Sciences that was staffed by Anne Keatley, now Solomon, called the Committee for Scholarly Communication with the People's Republic of China. It had been created by the energetic and farsighted Harrison Brown, foreign secretary of the NAS, and they had been preparing for such an opportunity for almost five years. Working with Anne and with the dozen or so subject experts on the White House OST staff, we were able to put together some forty different initiatives that could be suggested to the Chinese.[6]

Putting together the China science cooperation package was a memorable task. The results, though apparently modest, were a major breakthrough, although not codified in a formal agreement for another seven years. Working for Kissinger also proved to be memorable. Neureiter continued:

> Working for Henry Kissinger was a special experience. Typically, you would hand in a paper in the morning and at the end of the day, you were told that it was no good and should be redone. So you worked most of the night, turned in the new draft the next morning and then around 5:00 p.m. again heard that it

still needed work, et cetera. After a few such repeats, it did finally pass muster, and that was the last we ever heard of it.

We never got a report on how our document was used or discussed at the meetings in Beijing. But when we saw the Shanghai Communique, the formal document signed by both countries at the end of the visit, we felt that we had in fact had great success. Here is the relevant paragraph from the final official text: "The two sides agreed that it is desirable to broaden the understanding between the two peoples. To this end, they discussed specific areas in such fields as science, technology, culture, sports and journalism, in which people-to-people contacts and exchanges would be mutually beneficial. Each side undertakes to facilitate the further development of such contacts and exchanges." . . .

We considered those simple phrases to be a major victory and recognition of science cooperation as an effective instrument of constructive engagement. However, movement on both sides was quite slow until formal diplomatic relations were established on January 1, 1979, by President Carter and Deng Hsiao Ping.[7]

This idea of intergovernmental science and technology cooperation as a tool to repair relations with other countries became a pattern for the Nixon administration. The thinking behind it was much more instrumental and directly related to the US government's foreign policy than the exchange program initiatives of the Eisenhower administration, which emphasized direct people-to-people contact as a means of getting to know each other better. As discussed earlier, Kissinger more ambitiously described the basic rationale for such cooperation to the US Congress later in 1972 as something that would create "incentives for restraint."[8] This rationale was a significant step beyond the people-to-people vision articulated by President Eisenhower. It was to be explicitly political and linked in principle, moreover, to specific actions of the Soviet government. This was important and noted at the time. It soon became an object of controversy between the Congress and the administration, pitting scientific goals, on the one hand, against instrumental political ones, on the other. Over the years, this tension between science and foreign policy was played out in a number of different contexts. It remains one of the most fundamental issues, in my view, that all programs of government-sponsored scientific cooperation have struggled to address and must address in the future.

At the May 1972 Nixon-Brezhnev summit meeting, the very first bilateral agreement signed by the two leaders was not the science and technology agreement but the US-USSR Agreement on Cooperation in the Field of Environmental Protection, on May 23, just one day after the summit meeting got underway. This was quickly followed by the signature of the bilateral

Table 2.1. US-USSR Science and Technology Agreements

Field	Date Signed	US Executive Agency	USSR Executive Agency
environmental protection	May 23, 1972	Environmental Protection Agency	State Committee for Hydrometeorology and Control of the Natural Environment
medical science and public health	May 23, 1972	Department of Health, Education and Welfare	Ministry of Health
science and technology	May 24, 1972	Department of State	State Committee on Science and Technology
space	May 24, 1972	National Aeronautics and Space Administration	USSR Academy of Sciences
agriculture	June 19, 1973	Department of Agriculture	Ministry of Agriculture
transportation	June 19, 1973	Department of Transportation	Ministry of Transportation
studies of the world ocean	June 19, 1973	National Oceanic and Atmospheric Administration	State Committee on Science and Technology
peaceful uses of atomic energy	June 21, 1973	Atomic Energy Commission	State Committee for the Utilization of Atomic Energy
energy*	June 28, 1974		
artificial heart research and development	June 28, 1974	Department of Health, Education and Welfare	Ministry of Health
housing and other construction	June 28, 1974	Department of Housing and Urban Development	

*I have been unable to identify definitively the US and Soviet executive agents for this agreement. Its signature predated the creation of the first US civilian energy agency, the Energy Research and Development Administration (ERDA), which was only later combined with the Federal Energy Administration to create the US Department of Energy. The Soviet executive agent may have been the Ministry of Energy, but regrettably there are no documents I was able to find to verify either the US or Soviet responsible parties.

agreement on cooperation in medical science and public health (May 23, 1972); the agreement on scientific and technical cooperation (May 24, 1972); and the agreement on space (May 24, 1972).[9] These three executive agreements, which did not require Senate confirmation, were augmented over the next two years by eight more agreements, each tied to a mission agency in each country, to form a dense network of eleven formal programs (see table 2.1), funded and administered largely by the two governments.

Together, these agreements formally engaged eleven US government lead agencies and corresponding Soviet government ministries and state committees as executive agents responsible for the ensuring the funding and implementation of their associated activities, as well as a larger number of agencies and other government bodies in supporting roles. They engaged thousands of scientists from government laboratories, universities, and the private sector. Each agreement, with the exception of that for science and technology, was overseen by a bilateral joint committee chaired by the secretaries or ministers of the corresponding mission agencies; the mission agencies, in turn, were the executive agents of the respective agreements.

The science and technology agreement was an exception to this pattern. It was governed by a joint commission chaired by the science advisor to the president and the chairman of the USSR State Committee for Science and Technology (SCST). The NSF and the SCST were the agreement's US and Soviet executive agencies, although the US executive secretary of the science and technology agreement resided in the Department of State, while in contrast the executive secretaries of the other agreements were on the staff of the mission agencies responsible for them.

This anomaly was of concern to the Congress, because it seemed to be based on the primacy of politics, not science, in the programs. In 1976, the House Subcommittee on Domestic and International Scientific Planning and Analysis—the same subcommittee before which Kissinger had testified in 1972 when presenting his "incentives for restraint" rationale for the agreements—held new hearings on the state of the US-USSR intergovernmental agreements.[10] In a sign of irritation with Kissinger's instrumental political view of the programs, the subcommittee made it clear at the outset and in its key recommendation that the criteria for evaluating the exchanges must be scientific, not political:

> RECOMMENDATION 1: . . . The Cooperative Agreements in Science and Technology were entered into initially for the political purposes of reducing tensions and furthering international relations between the United States and the Soviet Union. However, their success both in terms of those objectives and in terms of furthering the application of science and technology to common problems in the two countries will ultimately depend on the benefits they yield to the United States and the Soviet Union. The agreements should therefore be evaluated on how well they serve to bring about and further an equal and reciprocal exchange of new and useful science and technology.[11]

On the scientific benefits, the subcommittee's members were skeptical but mostly open-minded. Rep. James Conlan (R-AZ) started the questioning by asking point-blank whether it was true that the Soviet joint commissioners were dictating the topics of cooperation. Presidential Science Advisor Guyford H. Stever's response was to cite the good will on both sides: "There is abundant evidence of good will and serious intent to engage in productive joint efforts. On the positive side, I can report modest but concrete results."[12] Yet despite this half-response, witnesses provided examples of clear mutual benefit and specific benefit to the United States. The star of the show was the Apollo-Soyuz Test Project. (This, however, was a bit of a sleight of hand because Apollo-Soyuz was never actually an integral part of the space agreement; indeed, it was not really a joint research project at all but a joint engineering feat. Yet it was very popular and well known to the public, and in congressional hearings, that's often what counts more than complete accuracy.) The subcommittee also cited three specific agreements as "promising" in terms of benefits: the science and technology agreement (especially the working groups on electrometallurgy and materials and on chemical catalysis), the world ocean agreement, and the environmental protection agreement. Access to previously unavailable Soviet technological advances (such as special metals and chemical processes) or valuable geophysical data (earthquakes, pollution, climate change) were the main positive examples cited in the committee report. Yet the Soviets' withholding of important data was also mentioned, for example regarding agricultural data. The witness from the US Department of Agriculture observed:

> The major problem, and the largest of all problems with the entire Agreement, is the failure so far of the Soviets to provide current situation and outlook data [i.e., forecast estimates] on production, utilization and foreign trade of major agricultural commodities. The Soviet refusal to provide the above data to the U.S. side has had the net effect of making the benefits obtained by the Soviets from the Agreement larger so far than those received by the United States, measured entirely in agricultural and agriculturally-related terms. Agreement activities have not yet proceeded far enough to make this imbalance unacceptable to the United States, but there is watchful concern.[13]

The statement from the US Department of Transportation also expressed reservations, in this case relating to the Soviets' inherent advantage in access to freely available information in the United States: "We have found that, despite their assertions to the contrary, the Soviets lag behind the U.S. in some areas where we had hoped to gain from their experience (e.g., magnetic levitation). . . . In order to maximize the cooperative relationship, we need more information concerning what the Soviets have to offer. In most cases we have to rely on what they say is available. Their task is easier because much of what is new here is covered in the open press."[14]

The subcommittee concluded that each working group's projects should be reviewed no less than annually and that "resources should be concentrated on projects which offer a demonstrable greater potential for more meaningful scientific or technical returns. Such evaluation will insure that worthwhile projects will not suffer at the expense of marginal projects."[15] Again, there was no mention here of any countervailing political benefits.

It is important to understand this point because in the intergovernmental agreements up until 1991, it was the congressional authorizations and appropriations of the *mission agencies* that were the engines that drove the programs' actual implementation. The State Department could say what it wanted about the overall, global balance of benefits, but when it came down to implementation, this was the business of the research and other executive branch agencies that had to justify the expenditures in competition with those for their core responsibilities. This was also a key reason why the subcommittee complained that the science and technology agreement's management was skewed, because this was the only case in which the executive agent was not the responsible mission agency, the NSF, but instead the State Department.

While it is very difficult to find accurate data about the levels of effort and funding in all these formal agreements, statistics carefully compiled by Catherine Ailes and Arthur Pardee, Jr. about activity under the science and technology agreement are indicative. They record that across that agreement's eleven working groups, 1,057 individuals from the United States alone took part in cooperative activities, including joint working group meetings, joint project-level meetings, workshops and seminars, research team exchanges, long-term individual visits, short-term individual visits, and US-side meetings. Of these, 48 percent were from universities, 18 percent from government, 19 percent from industry, and 14 percent from the nonprofit sector (1 percent were "affiliation unknown").[16]

Over the same period, the NSF, the main funder of the science and technology agreement on the US side, spent over $22 million on these activities—and this was just one agreement among eleven. Furthermore, the science and technology agreement was cut short after ten years while most of the others lasted for nearly twenty. My impression is that some of the other, agency mission–related agreements were even more active. These data should give a general feel for the level of activity in this new wave of cooperative programs. It is fair to say in broad terms that it was at least one order of magnitude higher than the volume of the Eisenhower-era exchange scientific programs on every scale.

The structure of the détente-era intergovernmental program also differed markedly from the exchange programs. New layers of formality and complexity appeared with the age of détente and the advent of the official intergovernmental cooperative programs of the 1970s and 1980s. While in the exchange programs, at least on the US side, the initiative to make visits bubbled up from the bottom, from individual scientists, in the intergovernmental programs individual initiative seemed to be stood on its head. Each intergovernmental agreement was chaired by a high-level joint committee or joint commission. These bodies, usually chaired by a mission agency head, authorized the establishment of working groups to address specific issues or research fields, and the working groups in turn outlined the types of projects that would take place and in many cases determine who should participate in them. Typically, joint meetings of bodies at each level required separate preparatory US-side (and presumably Soviet-side) meetings beforehand, not to mention interagency consultations to agree on the parameters for all the above. The result was a proliferation of meetings and bureaucracies to cope with this new, expanded, very hierarchical mode of expanded cooperation.

The content, as well as the form, of cooperative activities also looked quite different under the new intergovernmental programs. While in the exchange programs the typical activity was an individual scientist's visit for familiarization or research, in the intergovernmental programs, it was increasingly the group meeting. Joint committees and commissions, working groups, planning groups, and conferences, once brought into existence, all had to meet. To be sure, the scope and complexity of some of the new research efforts to be undertaken required such consultations, and so did the even more basic need to understand what each side had to offer in the agreed subject areas. At the same time, meetings are much more labor- and cost-intensive undertakings than individual visits, and this came to be reflected in the exponential growth of program funding alluded to earlier, as well as the bureaucratic support required. As a program manager, I was a beneficiary of all this superstructure—but not for long, because the entire science and technology agreement was terminated in 1982 to protest Soviet international behavior, as I will discuss in chapter 3. Indeed, the intergovernmental programs—in particular, the high-level science and technology agreement—were set up deliberately in such a way as to be available to "send messages" of displeasure when the occasion arose, as if that ever made any difference to the Soviet regime's conduct.

The detailed NSF-sponsored study of the US-USSR science and technology agreement cited earlier reveals that the profile of group versus individual activities taking place under that program included what must be regarded as an unusually large proportion of planning and other management meetings as opposed to research-oriented visits. Of the 410 activities recorded during this period, fully 37 percent of them (152 activities) were planning or other meetings at the working-group, project-group, or US-side level. To a certain extent, these meetings served the useful purpose of focusing effort on potentially productive and mutually beneficial topics of cooperation, as well as weeding out others that did not meet those criteria.[17]

In this way, the intergovernmental programs' more elaborate, hierarchical, and politically managed nature tended to generate weighty, formal, and costly superstructures within each area of activity that reflected and mimicked the hierarchy emanating from the top. As a whole, it was a showy, but very cumbersome and inefficient way to conduct scientific cooperation.

Notes

1. A unique and extremely detailed account of the development of the science and technology agreements of this period can be found in Ailes and Pardee 1986, 1–15.

2. As quoted in Ailes and Pardee 1986, 11. Kissinger's statement that "there were no cooperative efforts in science and technology" was obviously untrue on its face. The interacademy and scholarly exchanges had already been underway for some fifteen years. In this sense, Kissinger's statement appeared misleadingly to give the Nixon administration credit for inventing scientific cooperation with the Soviet Union. More importantly, however, until this time there had been no formal intergovernmental cooperation in science and technology, and this was an important departure, both in form and intent.

3. OST, together with the president's Science Advisory Committee was eliminated by President Richard Nixon in 1973 who opted not to appoint a replacement for its director, Edward E. David Jr., after he resigned. It was reestablished by the US Congress in 1976 as the Office of Science and Technology Policy (OSTP).

4. Interview with Norman Neureiter, January 11, 2016. He refers here to Reischauer's essay, "The Broken Dialogue with Japan," that appeared in the October 1960 issue of *Foreign Affairs*.

5. Neureiter interview.
6. Ibid.
7. Ibid.
8. Ailes and Pardee 1986, 11.

9. There is much rich detail on preliminary discussions about the science and technology agreement, which was really the centerpiece, within the Executive Office of the President (OST and the National Security Council) and the State Department in Ailes and Pardee 1986, 4–7.

10. *Review of U.S.-U.S.S.R. Cooperative Agreements on Science and Technology: Special Oversight Report No. 6* 1976.

11. Ibid., 8.

12. Ibid., 30.

13. Ibid., 33.

14. Ibid., 33–34.

15. Ibid., 10.

16. The working groups were Application of Computers to Management, Chemical Catalysis, Electrometallurgy and Materials, Microbiology, Physics, Science Policy, Corrosion, Heat and Mass Transfer, Earth Sciences, Polymer Sciences, and Scientific and Technical Information. Ailes and Pardee 1986, 41–72.

17. For example, in the field of microbiology, the US and Soviet sides were consistently at odds, with the US side insisting on basic research cooperation (where the Soviets, largely due to the ravages of Lysenkoism, were weak), while on the Soviet side there was a preference for what might be called "applied microbiology," which many in the United States considered shorthand for biological weapons. The discussion ended in a stalemate with no cooperative activities at all. On Lysekoism, see, for example, Graham 2016, Soyfer 2002, and Medvedev 1969.

3

SANCTIONS AND PERESTROIKA

OVER THE COURSE OF THE NEXT TWENTY YEARS, the détente-era cooperative "S&T" (science and technology) agreements, as they were collectively and informally called, served as an important vehicle for joint work in a wide variety of fields. Because of their formal, intergovernmental nature, however, they were also a barometer of the state of relations between the two superpowers. Almost from the beginning, there were serious problems that found their resonance in the agreements. Prominent among these were the treatment of Andrey Sakharov beginning in the early 1970s, which became a symbol of Soviet repression of dissent and human rights; the persecution of Jewish scientists who sought to immigrate to Israel and elsewhere, the so-called refuseniks, beginning in the mid-1970s; and aggressive Soviet behavior abroad from 1979 onward, including the war in Afghanistan, the imposition of martial law in Poland, and the shooting down of Korean Airlines Flight 007.

The bleakest time for bilateral cooperation, prior to the Crimea crisis of 2014, was from late 1979 through 1984. These dark days of bilateral science cooperation stemmed from a combination of private initiative on the part of the scientific community and US government action.

It was President Jimmy Carter, of course, who made human rights a pillar of US foreign policy. The first time, to my knowledge, that the US government used scientific cooperation as a stick to punish the Soviet Union was with regard to the human rights issue, specifically the trial of dissident Anatoly Shcharansky in 1978, when the United States postponed a meeting of the high-level bilateral Joint Commission for Science and Technology Cooperation in protest.[1] These sanctions, if that is what they can be called, were only temporary. Catherine Ailes and Arthur Pardee wrote:

> In July 1978, when the Soviet Union announced that the trial of Shcharansky would take place during the same week in which the U.S.-U.S.S.R. Joint

Commission on Science and Technology was scheduled to hold its annual meeting in Moscow, President Carter, as an expression of concern over the treatment of Soviet dissidents and the arrest of an American citizen in Moscow, indefinitely postponed the meeting of the Joint Commission and placed a freeze on most high-level official trips to the U.S.S.R. However, as relations between the United States and the U.S.S.R. improved, that moratorium quietly ended and the Joint Commission meeting was rescheduled.[2]

At the working level, cooperative activities, even under the joint commission's aegis, were never really interrupted by these events. A year later, however, there was a crisis of a different sort, this one originating at the grassroots level. The US-USSR Working Group on Special Topics in Physics under the science and technology agreements had met in Moscow in December 1978 to plan a joint working program. Co-chairing the group were David Pines of the University of Illinois at Urbana (and Los Alamos Scientific Laboratory) and Roald Z. Sagdeev, then director of the Soviet Academy of Sciences' Space Research Institute.[3] It was the coldest December in Moscow in memory, reaching negative forty degrees, the point where the Celsius and Fahrenheit scales cross; at the US embassy, a marine who raised the flag one morning lost his ears to frostbite.[4] The outside temperature matched the chill of the overall climate between the two countries. Nevertheless, in a cordial discussion the working group succeeded in agreeing to an ambitious and exciting plan of joint meetings engaging some of the leading theoretical physicists and astrophysicists from not only the two countries but also the world.

One of these scientists was Lev B. Okun', a senior scientist at the Institute of Theoretical and Experimental Physics, where he headed the Theoretical Physics Division. A particle physicist, Okun' was the first to introduce the concept of the hadron. He was an acknowledged world leader in his field, a tough and sometimes outspoken individual befitting of his prestigious status, and a Jew—but not a refusenik. To my knowledge, he had never been allowed to travel outside the socialist world. But the Soviet and US leaders of the physics working group agreed that he would be one of the Soviet participants in a joint working group meeting to be held the next July at the Aspen Physics Institute in Aspen, Colorado.

That, however, was not to be. As the workshop's date approached, the list of Soviet participants materialized from the Soviet Academy of Sciences, with one key name missing: Lev Okun', who will also reappear later in our story.[5] Cables of inquiry were sent from the National Academy of

Sciences (NAS), which managed the working group and where I worked as the coordinator, to the Soviet Academy of Sciences, but to no avail. Having consulted with the US working group members, at the last minute, an outraged Pines recommended to NAS president Philip Handler, who had led the NAS in adopting a strong position of protest regarding the Soviets' treatment of Sakharov, that the workshop be canceled. Handler agreed. The workshop was canceled, and the NAS disbanded the physics working group itself.

Thus ended a joint activity that had the potential to be the most scientifically meritorious, exciting, and prestigious effort under the intergovernmental science and technology agreement. All this transpired completely at the initiative and insistence of the US scientific community, with the US government not lifting a finger or saying a word. Some of my State Department friends gave me an "attaboy," but that was the extent of official involvement in this affair.

In this deteriorating climate, it was only seven months later that the US government imposed new sanctions, this time triggered by the Soviet invasion of Afghanistan in December 1979. In January 1980, Carter issued an ultimatum to Leonid Brezhnev that if the Soviets did not withdraw from Afghanistan in one month, the United States would boycott the 1980 Olympic Games to be held in Moscow. While it was not clear that he had the authority to do that, the boycott was upheld by the US Olympic Committee and joined by many, but by no means all, countries. But even in this case, the post-Afghanistan sanctions did not lead to a cessation of cooperative science activities. Instead, as before, they consisted of a ban on all high-level visits, planning meetings, and the like, but not on low-level cooperative science activities. Ailes and Pardee noted: "The Carter Administration moved quickly to defer scheduled high-level meetings to be held under the auspices of the U.S.-U.S.S.R. bilateral agreements in science and technology. All joint activities under the S&T Agreement were subsequently reviewed by the Department of State on a case-by-case basis. This was in reality a return to the *status quo ante*, since all activities had been routinely reviewed on a case-by-case basis.[6]

The programs for which I was responsible at the National Science Foundation (NSF) under the bilateral science and technology agreement continued,[7] as did prominent nongovernmental programs such as the interacademy exchanges. Indeed, the generous congressional appropriation (over $3 million) for these NSF-managed cooperative programs continued and was renewed later that year. Still, the Afghanistan sanctions were the first sign that even scientific cooperation, which according to many in the scientific community did not know national borders, was not in a protected category as far as government-sponsored programs were concerned.

Thus the protestations of the Carter administration over the Shcharansky trial, the persecution of Sakharov, and the Soviet invasion of Afghanistan, despite all the public uproar and symbolic actions like temporary bans on high-level meetings and visits, had little tangible effect in my corner of the world. I was left to wonder what it would take to trigger serious sanctions that would demonstrate that the formal intergovernmental programs were, as Henry Kissinger testified in 1972, "incentives for restraint."[8]

Ultimately, it was left to the Reagan administration to impose sanctions with sticking power, although in a very inconsistent way. In December 1981, the Soviet Union imposed martial law in Poland in an effort to quash the workers' dissident movement led by Lech Wałęsa and increasingly supported by the Catholic Church, in the person of Cardinal Karol Wojtiła, the future Pope John Paul II. As a result, the Reagan administration decided to cancel the scheduled renewal of all formal intergovernmental science and technology agreements (and perhaps others) that were due to be renewed in 1982. This included the agreements on science and technology, energy, and space.

Interestingly, however, two other agreements that were up for renewal in 1982—medical science and public health, and environmental protection—were allowed to be renewed automatically. Evidently, these agreements were in a more protected category because of their more benign areas of application. The other three less fortunate agreements were more vulnerable, for various reasons.

The science and technology agreement was always vulnerable because of the perception of many in the US government, in both the executive and congressional branches, that it represented a net loss to US science and a possible backdoor to sensitive technologies for the Soviets. The intergovernmental agreement in space, counterintuitively, was expendable because most of the spectacular space cooperation between the two countries, such as Apollo-Soyuz, occurred independently of these high-level, politically motivated bilateral agreements. With regard to the energy agreement, it had likely gone out of favor in the Reagan administration for domestic political reasons, because the administration was hostile in principle to a broad government role in sponsoring new research, especially on renewable sources;

the energy agreement, too, was distinct from the Agreement on Peaceful Uses of Atomic Energy, which was clearly more germane to the bilateral relationship and in any event not due for renewal in 1982. In addition, in 1983 the United States decided not to extend the agreement on cooperation in transportation in response to the September 1, 1983, Soviet shooting of Korean Airlines Flight 007 over Sakhalin Island. "By the end of 1983," according to Yale Richmond, "the number of persons exchanged under the eleven cooperative agreements had fallen to about 20 percent of the 1979 level."[9]

John Zimmerman, a Foreign Service officer who was working at the time on the Soviet desk at the State Department, remembered that time well:

> Right after that [the Soviet imposition of martial law in Poland in December 1981], the White House canceled all of the bilateral programs with the Soviet Union. So when I arrived on the Soviet desk as the new officer focusing on scientific and cultural exchanges, I spent a great deal of time picking up the pieces of that decision, which included negotiating the return of a major item of equipment sent to the USSR under the Energy Agreement. In return for sending back this equipment—a large magnet to be used for joint magnetohydrodynamic experiments—the Soviets wanted consumer technology items (e.g., VCRs and home movie cameras). Although it sounds simple, these negotiations kept me and the interagency community quite busy.[10]

This, however, was not the end of the curious story of the Reagan administration and scientific cooperation with the Soviet Union. On March 8, 1983, Reagan famously pronounced the Soviet Union to be an "evil empire."[11] But just months later—and this was well before Mikhail Gorbachev came to power as Communist Party of the Soviet Union (CPSU) general secretary in 1985—officials in the administration were pondering ways to restore the relationship. There is even evidence that Reagan himself was having a change of heart. Zimmerman said:

> In August 1983, the Soviets shot down the Korean airliner flight KAL 007, an event I remember clearly because all of us on the Soviet desk were manning the Operations Center and seemed to be working twenty-four hours a day. In any case, sometime in late 1983 or early 1984, I decided that if we were ever going to warm up our bilateral relationship, the S&T side might be the place to start, using the canary-in-the-coal-mine logic that if we couldn't revive this aspect, what hope would there be for other, more complex issues. So, I wrote a series of memos stating in essence, that if the White House is interested in sending a signal to the Soviets or in actually reviving at least this part of our relations, here are some things we could do....

> The head of the Soviet desk, Thomas Simons, worked with Ambassador Jack Matlock, the National Security Council [NSC], Deputy Assistant Secretary Mark Palmer, and Assistant Secretary Richard Burt to push my memos up the chain of command. Some of my ideas . . . were shot down rather quickly; the ones, however, which dealt with the S&T exchange programs were well received and the message came back to the effect, "Send us more."
>
> As I recall, I proposed positive steps for all the agreements. This initiative was handled very, very quietly, but seemed to be aided by what I understand was a desire by the First Lady and NSC advisor Bud McFarlane to show that the administration was making at least some progress in working with the Soviets. The entire process was capped off by a speech which President Reagan gave in June 1984, announcing a reinvigoration of US-Soviet relations in the S&T area. Some of my language from the first draft of his speech survived the editing process, but the real pleasure was knowing that when the conditions were right, a simple desk officer could actually have an impact on US foreign policy.[12]

Indeed, this was one of those unique historical moments, though few recognized it as such, when personalities and politics intersect most powerfully. Reagan, despite his tough public rhetoric, was actually having a change of heart. "This was a time," Zimmerman continued, "when the president himself was signaling that the atmosphere in bilateral relations was beginning to shift":

> Reagan gave a televised speech on arms control, the draft of which had circulated around the Soviet desk. As I recall, Tom Simons and Mark Palmer were the original drafters. However, when the president actually gave the speech, he added a paragraph apparently in his own handwriting which was a short story about an American and a Soviet couple who have a chance encounter at a bus stop. In his telling, the president had the two couples seeking shelter from the rain. . . . The scenario was a tad sappy, but his message was clear: in his view, on a personal level we are all united by a common humanity in spite of the differences in our political systems. That said, my colleagues and I were shocked that the president would present such a "forward-leaning" view of the Soviets: certainly, none of us would have the temerity to insert something of this nature into his speech.[13]

Here were Reagan's actual words during this speech, which was broadcast from the East Room of the White House on January 16, 1984, just a few months after the KAL 007 incident:

> More than 20 years ago, President Kennedy defined an approach that is as valid today as when he announced it. "So let us not be blind to our differences,'" he said, "but let us also direct attention to our common interests and to the means by which those differences can be resolved."

Well, those differences are differences in governmental structure and philosophy. The common interests have to do with the things of everyday life for people everywhere. Just suppose with me for a moment that an Ivan and an Anya could find themselves, oh, say, in a waiting room, or sharing a shelter from the rain or a storm with a Jim and Sally, and there was no language barrier to keep them from getting acquainted. Would they then debate the differences between their respective governments? Or would they find themselves comparing notes about their children and what each other did for a living?

Before they parted company, they would probably have touched on ambitions and hobbies and what they wanted for their children and problems of making ends meet. And as they went their separate ways, maybe Anya would be saying to Ivan, "Wasn't she nice? She also teaches music." Or Jim would be telling Sally what Ivan did or didn't like about his boss. They might even have decided they were all going to get together for dinner some evening soon. Above all, they would have proven that people don't make wars.

People want to raise their children in a world without fear and without war. They want to have some of the good things over and above bare subsistence that make life worth living. They want to work at some craft, trade, or profession that gives them satisfaction and a sense of worth. Their common interests cross all borders.[14]

Reagan's biographer, H. W. Brands, offers authoritative confirmation of Zimmerman's impressions of the president's position on Russia. It was hardly the categorically negative picture that was widely depicted at the time. "Reagan wasn't calculating," he wrote, "but he was canny." In other words, Reagan, while sticking to his tough public profile on the Soviet Union, was sensing that a time for change might be on the horizon. According to Brands:

> Other administrations have played the bad-cop, good-cop routine by assigning the separate roles to different actors. . . . Reagan played both sides himself; one in words, the other in deeds. He thundered against Soviet perfidy for the television camera, and he meant everything he said. But the substantive measures he took against the Soviets were remarkably modest. He suspended American landing privileges for Aeroflot, and he postponed negotiations toward some bilateral Soviet-American agreements. But otherwise it was business as usual.
>
> The result was scarcely short of brilliant. Reagan credibly reiterated America's claim to the moral high ground vis-à-vis the Soviets, benefiting himself politically and America in the eyes of the part of the world he cared about. But to Soviet leaders, who, he always contended, paid more attention to deeds than to words, he left open the door for future negotiations.[15]

So while Reagan talked tough about the "evil empire," in fact he acted quite differently, apparently because he dearly wanted to bring the Soviets to the negotiating table to discuss arms control. Then in 1985, after he met Soviet CPSU general secretary Gorbachev at the Reykjavik Summit

Meeting, the atmosphere began to thaw once again. Scientific cooperation continued apace in the remaining extant bilateral programs. Even in agencies whose special programs and associated special funding had been cut off, such as the NSF, collaboration by individual scientists, often with funding through their standard research grants, continued unaffected.

By 1987, the political atmosphere had improved sufficiently that it was even possible for the NSF and the State Department to initiate discussions of negotiating a new intergovernmental agreement on basic scientific research to replace the 1972 science and technology agreement. As an NSF senior program manager, I was personally involved in the sixty-two interagency meetings, chaired by the Office of Science and Technology Policy (OSTP), that were needed to obtain consensus on the US side, and in all the negotiations with the Soviet side on the agreement. There were some moments of suspense over the agreement's signing due to the last-minute objections of the president's science advisor at the time, William R. Graham, who tried to sabotage the agreement on the day the agreement was to be initialed. Graham insisted on reviving an issue that had been debated at length by the two sides during the yearlong negotiations—the definition of "fundamental research"—and that had been formally resolved, or so we thought.

This last-minute standoff within the US administration resulted in a peremptory summons to delegation leader Richard J. Smith, then the State Department acting assistant secretary for oceans and international environmental and scientific affairs, to appear at the National Security Council. After hearing and rejecting arguments by OSTP staff and the Office of the US Trade Representative, Smith took the courageous step of initialing the agreement anyway.[16] Several days later, he recalled, Secretary of State George Shultz wrote to him that "the fact that some of the agencies later backed out after they had already given their consent [to initialing the text] is their problem, not ours."[17] The new basic science agreement was eventually signed by the two foreign ministers—and I was back in business at the NSF in managing US-Soviet science cooperation. I do not know how often this kind of tumult occurs in international negotiations, but having been involved in many of them with the Soviet Union, I had never before witnessed such sorry disarray.

Similarly, in April 1987 the two governments agreed on the new bilateral Agreement on Cooperation in the Exploration and Uses of Outer Space for Peaceful Purposes, filling the gap in that area that had been created by the cancellation of the 1972 Agreement on Space Cooperation that had been signed at the Nixon-Brezhnev summit and canceled in 1982 by the Reagan administration. As Susan Eisenhower described in her account of US-Russia space cooperation, by this time, relations had thawed sufficiently to put space cooperation back on the table.[18]

In these tumultuous years, then, it was foreign policy that drove down formal science cooperation programs when things got worse—and the same programs that sometimes served as bellwethers of improved relations not long after. In any case, it does not appear, that they constituted "incentives for restraint"[19] in Soviet behavior, as Henry Kissinger claimed they would before the Congress in 1972, in any sense of the term. When the programs were used as sticks to punish bad behavior, they were little more than slaps on the wrist. When they reappeared as relations improved, it was perceptions of improved Soviet behavior—and indeed, under Gorbachev, of the prospect of reform or imminent collapse—that preceded and enabled the programs to reappear, not the other way around. Although nobody noticed, the Kissinger doctrine had been thoroughly discredited, even though the notion continued to live its own life in the minds of policymakers, legislators, diplomats, and others. This disconnect between the often overblown political expectations about scientific cooperation and its actual impact is one of the greatest occupational hazards I encountered during my years of work in this area, both in and outside of government. It did not start with Kissinger and it certainly did not end with him, but his service was to give it its clearest formulation.

Notes

1. Shcharansky changed his name to Natan Sharansky after immigrating to Israel.
2. Ailes and Pardee 1986, 34. The American who was arrested was Francis Jay Crawford, an employee of International Harvester Company. He was charged by Soviet authorities with currency violations. This action was a much-expected retaliation that had even been threatened in general terms by Soviet foreign minister Andrey Gromyko after the United States arrested two Soviet United Nations employees in New Jersey on suspicion of espionage, See Maggs 1986 and Gwertzman 1978.
3. Sagdeev later immigrated to the United States in the late 1980s after marrying Susan Eisenhower, a granddaughter of President Dwight Eisenhower, and became associated with the University of Maryland in College Park.
4. I traveled to Moscow with the physics working group as the US administrative coordinator. I had never before had the experience of long icicles forming on my nose and hopefully never will again.
5. See the section in chapter 4 titled, "A Letter Sets Off an Avalanche."

6. Ailes and Pardee 1986, 34.
7. I moved from the NAS to the NSF's Division of International Programs in October 1979.
8. Ailes and Pardee 1986, 11.
9. Richmond 1987, 78.
10. Interview with John Zimmerman, September 30, 2015.
11. Reagan 1983.
12. Zimmerman interview.
13. Ibid.
14. Reagan 1984.
15. Brands 2015, 423.
16. The entire episode is documented in Smith 2009, 81–84.
17. Ibid, 84. The brackets are those inserted by Smith.
18. S. Eisenhower 2004, 18–20. See also Albrecht 2011, 89–91.
19. Ailes and Pardee 1986, 11.

4

AFTER THE FALL
New Times, New Approaches

On December 26, 1991, the Union of Soviet Socialist Republics formally came to a quiet end, marking what Russian president Vladimir Putin has called "the greatest geopolitical catastrophe of the twentieth century."[1] The Soviet Union's sudden disappearance into the dustbin of history shook science and technology in Russia and the other constituent union-republics to the core.[2] The massive Soviet scientific community, the largest one in the world by any standard,[3] with over 1.5 million scientists and engineers in 1991,[4] with a sprawling institutional structure of thousands of scientific research institutes belonging to government ministries and academies of sciences, was suddenly sliced into sixteen separate pieces.

When the Soviet Union collapsed in 1991, this system came to an abrupt end. Military funding for research, its mainstay for decades, almost completely dried up overnight, leaving hundreds of thousands of scientists at loose ends. The choices were grim. To remain in science, one had to figure out how to live on virtually zero income (in many cases, not more than the equivalent of twenty dollars a month) and a barter economy—or to emigrate and join the "brain drain" that had been underway for many years, especially for scientists of Jewish origin. Many, indeed, took the latter course, but in perspective they were the fortunate few who already had managed to build strong professional contacts abroad. Many more scientists, however, took the route of "internal emigration"—abandoning science entirely. Some were able to translate their mathematical talents into the banking and finance industries for relatively well-paying jobs, but the overwhelming majority took menial positions driving taxis and buses, guarding warehouses, and the like.

The crisis affected institutions as well as people. Entire sections of research institutes previously packed with scientists and technicians busy at research became ghost towns, though with time some institutes caught on that they could rent this empty space to commercial ventures as a source of income—a clever practice driven by necessity that turned out to be a significant factor in the academy's undoing two decades later. That the institutes and their scientists themselves could become entrepreneurs and quickly redirect their work toward making money, however, was another matter entirely. In fact, it was quite impossible. The experience and the talents simply were not there, having been suppressed by the Soviet command-economy, top-down, military-driven scientific research system.[5]

Broke and broken, the Russian (and former Soviet) science community was in desperate straits. Western governments worried that scientists who had worked in secret institutes on weapons of mass destruction (WMDs)—nuclear, chemical, and biological—might be willing to sell their services to buyers from rogue states or terrorist groups. For some there was also a moral and cultural issue, that one of the great treasures of Russian history—the intelligentsia—faced sudden extinction. It seemed to many, both within Russia and abroad, that such people and attitudes could provide a core of support for a new Russia as a democratic society. As it turned out, though, the scientific community's sudden impoverishment caused by the Soviet Union's collapse, the disappearance of ideological compulsion, and the replacement of political dictatorship with the rude appearance of the laws of the market all but destroyed any notion of privilege or noble mission in this real of society. It was humiliating and demeaning. The main point, now, was simply to survive.

At the same time, dramatic changes were about to take place in Western countries' strategies on cooperation in science, technology, and other fields. From a formidable rival, the states of the former Soviet Union (FSU) overnight became objects of assistance. The motivations behind this shift were complex and often disparate or even contradictory. One concern, as noted above, was about the dissolution of a great scientific establishment. Another was the dreaded threat of "brain drain," in two dimensions: In the civilian sector, a mass exodus of ex-Soviet scientists flooding our campuses and competing for scarce jobs with American (and other foreign) scientists; and the even more problematic issue of highly talented WMD scientists selling themselves to the highest bidder. In both cases, many thought that it would be preferable, and important to global security, to help ameliorate their situation at home. Finally, in the Western euphoria over the fall of communism, there was a sense that the West, and particularly the United States, could recreate the newly independent countries in our own image, imbuing them with market economies (including the growth of "knowledge economies" based on technological innovation) and democratic institutions. As Stephen F. Cohen has argued, this fatuous assumption was probably doomed from the start.[6]

In short, the quiet political revolution of the fall of communism was accompanied by a revolution in the decades-old approach of Western countries, and in particular the United States, regarding cooperation with this part of the world. This chapter is the story of the emergence of new and previously unimaginable thinking about the goals and methods of scientific cooperation between the United States and the newly independent states of the FSU, and the surprising forms they took.

Bilateral Intergovernmental Programs Continue in a New Guise

One of the first, and perhaps least interesting in a historical sense, developments in bilateral science cooperation after 1991 were efforts to keep the framework of formal intergovernmental agreements in place—at least between the United States and Russia. The bilateral US-Soviet intergovernmental programs initiated in the early 1970s during the Nixon-Kissinger détente era reemerged as agreements between the United States and the Russian Federation (and, in some cases, with the governments of other successor countries of the region). The network of this earlier layer of programs remained extensive, as follows:

- Agreement on Science and Technology Cooperation (December 16, 1993)
- Agreement on Cooperation in the Fields of Public Health and Biomedical Research (January 1994)
- Agreement on Cooperation in Research on Radiation Effects for the Purpose of Minimizing the Consequences of Radioactive Contamination on Health and the Environment (January 1994)
- Agreement on Cooperation in the Field of Protection of the Environment and Natural Resources (June 1994)
- Agreement on Scientific and Technical Cooperation in the Fields of Fuel and Energy (June 1992)
- Agreement Concerning Cooperation in the Exploration and Use of Outer Space for Peaceful Purposes (1992)
- Agreement on Scientific and Technical Cooperation in the Field of Peaceful Uses of Atomic Energy
- Agreement on Cooperation in Ocean Studies (1990)[7]

These intergovernmental programs came under the relatively loose umbrella of the Gore-Chernomyrdin Commission, which lasted until the end of the Clinton administration and the accession of Vladimir Putin to the Russian presidency. Activities under some agreements continued more or less apace, as in the oceans studies agreement, where the research agenda was linked to large-scale multinational programs. In other areas, for example the science and technology agreement, the formal agreement provided a basis for new relationships, such as that between the US National Science Foundation (NSF) and the newly established Russian Foundation for Basic Research. In one case, however—the Agreement on Peaceful Uses of Atomic Energy—everything came to a halt because the two governments could not agree on the issue of how to handle liability for accidents or other damages for several years. This difficulty was a source of mystification to many, but one supposed that deeper issues were in play.

The interacademy exchange programs—the core vehicle for individual scientists go back and forth during the Cold War era—also continued after the breakup for another eighteen years. But in the new, more open atmosphere of the 1990s and early 2000s, the need for structured programs with formal sponsorship for individual scientific visits had faded and came to be replaced with direct support through academic appointments, much as with the rest of the world. By 2009, the interacademy programs came to an end—just at the time, some might argue, when the xenophobia and repression characteristic of the Cold War years that gave birth to the venerable interacademy exchange program were returning to Russia.

In introducing this section, I suggested that the continuation of the intergovernmental agreements, while significant, were the least interesting development in bilateral science cooperation in the post-1991 period. And this was so. Everywhere else one looked, everything else was stood on its head. Underlying philosophies and goals, program management, the debut of major third-party programs, funding sources and funding flows—all exploded the carefully designed models of the Cold War years. In the words of one State Department science officer who had been deeply engaged in both periods, "all hell broke loose."[8]

A Letter Sets Off an Avalanche

The history of the Western response to the post-1991 crisis in FSU science begins with a desperate appeal that appeared in a little-noted letter from CERN (the European Organization for Nuclear Research) Director General

Carlo Rubbia to the president of France, François Mitterand. In terms of the history of international scientific cooperation, it is perhaps the single most financially productive letter that has ever been written.

Even before the end of December 1991 and the collapse of the Soviet Union along with its system of lavish military funding of scientific research, Soviet scientists saw the writing on the wall. The looming crisis was clear to all.

These scientists included a group of Soviet physicists working at CERN, led by the distinguished Russian theoretical physicist, the late Lev Okun'. In August 1991, Okun' drafted the dramatic "Appeal for the Salvation of Basic Science in Russia,"[9] calling for creation of an enormous emergency fund, which the appeal estimated was what would be needed to support basic research in Russia in 1992. It further specified that the funds should be made available not in the traditional top-down manner through institutes, but instead through project-oriented funding to individual researchers on the basis of merit. Rubbia sent the appeal[10] on to Mitterand in a letter dated September 26, 1991, who endorsed it a week later, "though the matter," according to a history of the period by Irina Dezhina, "was then stalled in governmental bureaucratic structures."[11]

In his letter, Rubbia pointed out that "the contributions of Russian science are part of the cultural patrimony of Europe and even of the whole of humanity.... The traditions of Mendeleev, Pavlov, Lobachevski, Kapitsa, Landau, Sakharov, and so many others are today continued through brilliant schools in mathematics, physics, astronomy, chemistry, and biology. Their disappearance would be a very grave loss, not only for Russia, but for world science and culture."[12]

To remedy this dire situation, Rubbia continued, "it is a matter of creating an International Foundation that would fund Russian science in its best fields and that would organically ensure its fruitful collaboration with Western science."[13] This was the first mention of a new body—an independent, international foundation—to provide emergency support for the best of Russian science. The word *Russian* rather than phrases such as "of the former Soviet Union" does strike a somewhat jarring tone today, but it must be kept in mind that the Russian group at CERN was, after all, Russian, and that in September 1991, the future of the USSR's geographical components was totally unknown and unknowable.

Although Rubbia's letter did not mention a specific sum of money, I do distinctly recall at the time seeing a separate document, which I believe was the text of the Russian CERN group's original appeal that was sent to Rubbia, specifying a sum of $100 million. As Dezhina wrote: "This marked the first appearance of the magic number of $100 million, which, according to many of those who witnessed the processes that transpired, subsequently reappeared frequently during discussions of aid programs for scientists of the former Soviet Union."[14] We will encounter this number, along with other previously improbable large numbers of dollars in support of this effort, throughout the rest of this chapter. The Rubbia-Okun' initiative (to which I refer hereinafter as "Rubbia-Okun'") was the visionary historic spark that set in motion many times that amount of support for former Soviet scientists over the next twenty years.

Even more important than funding or activity levels, and driving them both, was a marked shift in the thinking of many about the purpose of scientific cooperation with Russia. This is exceedingly important to understand, because in part it explains, in my view, the fate of some of these programs as time went on. In a sense the tone was set by the Rubbia-Okun' appeal of September 1991, even before the Soviet Union met its end. The appeal was for emergency assistance "*for the salvation of basic science in Russia* [emphasis added]." Thus began the notion that it was in the West's interest not only to cooperate with science in Russia,[15] but to come to its aid. As we shall see, this idea was responsible for some of the most spectacular successes, as well as failures, in science relations in the post-Soviet era.

A Fundamental Reorientation of US Policy

While these private appeals were percolating in the international science community, the US government was conducting its own policy reassessment. In early 1992, D. Allan Bromley, President Reagan's science and technology advisor, commissioned a study by the National Academy of Sciences (NAS) "to consider how to preserve the basic science capability of the former Soviet Union (FSU)." The resulting report, based on the conclusions of some 120 leading US scientists and engineers at a one-day workshop on March 3, recommended a series of measures aimed at support of FSU science for two key purposes: to build democracy and to prevent "brain drain." The key text in the main letter-report addressed to Bromley read as follows:

> Scientists and engineers in the FSU will play a key role in the economic revitalization necessary for a successful transition to open and stable market-driven democratic societies, which is in the economic and security interests of the U.S. New scientific and technological challenges in civilian areas for

FSU specialists can help divert technical talent away from military pursuits. Achieving U.S. goals of shrinking and redirecting the FSU military R&D effort and developing the S&T component of the civilian economy will require providing new opportunities for both FSU weapon scientists and non-weapon scientists, especially in collaborations with American scientists. These opportunities for collaborative efforts will also allow FSU specialists to help expand frontiers of knowledge in areas of direct interest to the American scientific community and to U.S. business.[16]

This set of objectives was a radical departure from the previous thirty-five-year history of science cooperation with the FSU. In fairness, the NAS report did not suggest substituting these new objectives for the traditional formula of mutually beneficial scientific cooperation; instead, it was an overlay. But as an overlay, it was a powerful reformulation of US interests and policy in this area. While it recommended budget increases for all programs, including the traditional cooperative programs, the report recommended that by far the bulk of the weight of any new funds should fall on programs for nonproliferation and assistance.[17] In practice, this was where just about all of the new funds that were actually available were directed in coming years.[18]

The Lab Scientists and the Scientific Societies

While policymakers were considering these strategic and tactical issues, it was individual scientists and their professional societies, not government, who took the first steps. To my knowledge, the first to respond were the nuclear physicists from the Department of Energy's Lawrence Livermore National Laboratory in California and the Los Alamos National Laboratory in New Mexico.[19] Livermore and Los Alamos, as the United States' leading nuclear research and nuclear weapons laboratories, already had informal ties and contacts with some Soviet counterparts in both academy and Ministry of Atomic Energy research institutes and even in the so-called closed cities.

The latter were secret military facilities whose existence was not officially recognized and whose identities were code-named for their distance from nearby population centers, such as "Chelyabinsk-72" and "Arzamas-16"; it was here that the most sensitive basic and applied research and development were performed on nuclear weapons. It is also likely that there were individual ties between Okun' and his Russian colleagues at CERN and the US labs. Be that as it may, the Livermore scientists answered the call by visiting Russian colleagues as early as December 1991 and delivering cash assistance from their personal resources.

The lab scientists soon thereafter became the semi-official vanguard of US government efforts to reach out to the FSU—in particular, Russia—rooted in deep concern about the security of its nuclear weapons arsenal. Siegfried Hecker, at the time the director of Los Alamos, related:

> What did it for us was that finally, in December 1991, Admiral Watkins, who was secretary of energy at the time, came back from a cabinet meeting with President George H. W. Bush and said, "The president laid out his concerns about the potential brain drain of the Russian nuclear weapon scientists as the Soviet Union came apart." He continued, "The president wants to know what are we going to do about it."
>
> I had been trying to get over there for more than a year and I said, "Look, Admiral Watkins, why don't we go ask them? We know their people. I'm lab director over here. You know, if I had my people in trouble, I'd have some ideas. So why don't we go ask them and see what their ideas are?"
>
> This was December 16, 1991, and Watkins said, "Go! Before Christmas!"[20]

The relationships between the United States and Russian weapons scientists that Hecker, together with Livermore director John Nuckolls, sought to foster through this visit developed into what widely came to be called the "lab-to-lab" programs. They played an extremely important, semi-independent role as a channel for both scientific and security-related cooperation in the coming years, as will be discussed in this and other chapters.

The next to respond were the scientific societies.[21] In particular, the American Physical Society (APS), the American Astronomical Society (AAS), and the American Mathematical Society (AMS) raised substantial sums of money from their dues, direct donations from their members, and contributions from external sources, including the Alfred P. Sloan Foundation, the NSF, and George Soros. Dezhina reports that in this manner, the APS collected $1.3 million, the AAS some $400,000, and the AMS also about $400,000, in support of these efforts. The societies set up national and international committees of scientists to receive proposals for support, evaluated them on a competitive basis, and awarded individual grants in the range of $30 to $100 a month to thousands of their former Soviet colleagues. They also used these funds for a variety of other purposes, such as purchases of computers and distribution of journals.

This unprecedented work by American scientific societies was historically noteworthy not only because of the money it raised and the benefits it conveyed to their peers in the FSU but also because of how it was done. Prior to this time, the concept of a competitive grant had no meaning in the former Soviet scientific space, as research funds had always been distributed in a top-down manner. Beginning with the societies' efforts, the very word *грант* (literally, "grant") made its appearance in the Russian language. Secondly, the donors went to great lengths to ensure that the funds' impact would not be diminished by taxes or corruption. They obtained the Russian government's agreement, at first informally and then through a formal exception in the tax code, to exempt foreign grants for scientific research from taxation. This tax loophole, originating in 1992 to facilitate some $2 million in emergency grants from the societies, eventually became the sine qua non for all science assistance efforts for the next twenty years, the bedrock of international and multinational programs that brought billions of dollars of support to scientists in Russia and elsewhere in the region. It started here.

Finally, simply getting the money from point A to point B in these chaotic early years was far from a straightforward matter. It was clear, first, that the funds could not be channeled through the Russian research institutes, both because of formal tax withholding procedures and because of well-founded fears of corruption. In the very beginning, many American scientists from the societies—and even some, according to urban legend, from the US national labs—literally bootlegged cash in denominations of twenty-dollar bills in their socks. (The APS had actually managed to set up a formal banking procedure.)[22] With time, the larger and more formal programs that subsequently emerged developed their own carefully managed administrative systems, in cooperation with selected local Russian banks, to deliver funds directly to their individual recipients. The societies and others then often piggybacked on these channels to transfer funds and other types of support.

It is impossible to overestimate the historical importance of the scientific societies' pioneering efforts in laying the groundwork for the far more massive programs of assistance and cooperation that followed them. They were the first to establish and implement the concept of competitive, investigator-initiated grant funding of science in the countries of the FSU; they pried open a tiny tax loophole for foreign scientific research grants that grew into a floodgate for massive infusions of funds that they could never have foreseen; and they dramatically illustrated the need to establish formal financial and administrative structures to deliver such support in a reliable, accountable, and transparent manner.

In parallel with the individual and group initiatives described above, discussion continued around the seminal concept advanced by Rubbia and

Okun' for a massive, $100 million fund to assist Russian scientists. By early 1992, the concept had begun to subdivide into at least four separate streams, each with a label ranging from $25 million to $100 million.

Chronologically, the first practical outcome of Rubbia-Okun' occurred in February 1992 in a meeting between US Secretary of State James Baker and Russian Federation President Boris Yel'tsin. Concerned about the leakage of Russian military secrets to countries of "the third world,"[23] the United States proposed creation of a multilateral body with contributions from the United States ($25 million), the European Union ($29 million), and Japan ($17 million). Two years later, in 1994, the International Science and Technology Center (ISTC), an international organization, would arise on the basis of this discussion.

It is important to note here that Rubbia-Okun' did not refer in any way to military or nuclear research of any kind; it was explicitly oriented toward the kind of basic scientific research typically funded by the NSF in the United States rather than more "applied" research. However, there were close relationships between the theoretical physicists in both the basic and defense-oriented research communities in both Russia and the United States, which was a particularly active conduit for information and concern. It is thus reasonable to suppose, and it was certainly my impression at the time, that the notion of massive Western support to ameliorate the various crises arising from the collapse of science in the FSU was widespread in government as well as academia.

The second subdivision of Rubbia-Okun' began in the same month, in February 1992, with the visit of Russian Academy of Sciences president Yuriy Osipov to the office of US representative George Brown of California. Brown, chairman of the House Science and Technology Committee, was an avid advocate of international scientific cooperation. Osipov appealed to him for the US government to set up an assistance fund. A consummate and effective legislator, Brown introduced HR 4550 on March 24 authorizing the establishment of "an endowed, nongovernmental, nonprofit foundation to encourage and fund collaborative research and development projects between the United States and Russia, Ukraine, Belarus, and other democratic republics emerging from the former Soviet Union."[24] The bill called for appropriations of $50 million each in fiscal years 1992, 1993, 1994, and 1995. Thus was born a second concrete proposal for the embodiment of the Okun'-Rubbia fund. From this legislation, a year later, sprang the US Civilian Research and Development Foundation for the Independent States of the Former Soviet Union, or CRDF, now called CRDF Global.

Next, in the summer of 1992, responding directly to the letter addressed to him by Rubbia and Okun', French president Mitterand confirmed his country's contribution of $27 million[25] to a new international organization, which, when joined by other EU countries, became the International Association for the Promotion of Cooperation with the Scientists of the Former Soviet Union, or INTAS.[26]

The final and fourth product of the Rubbia-Okun' vision was announced in December 1992 by the American philanthropist and financier George Soros. Soros had recently made a killing of the pound sterling in September, netting nearly $1 billion, and he decided to dedicate a portion of the proceeds to a fund to assist science and scientists in the FSU. This was the origin of the International Science Foundation (ISF), which was initially pegged at $100 million and which over its brief lifetime (1993–96) provided over $140 million in grants and infrastructural support for science in Russia (about 80 percent of its work) and the other countries of the FSU.

If we simply run a total of the raw numbers mentioned in this section, we see that over the course of just one year from the appearance of the seminal letter from Rubbia and Okun' calling for a $100 million fund, commitments or formal legislative authorization of almost four times that amount appeared, virtually overnight, to address the crisis of science and scientists triggered by the fall of the Soviet Union. All these programs were actually implemented, though some (such as CRDF) not at the levels originally anticipated; however, over time the cumulative effort was immense and unprecedented.

The number and nature of institutional actors also changed dramatically. While in the Cold War period, scientific relations between the United States and the Soviet Union were dominated by one semiprivate institution—the National Academy of Sciences—and while in the détente period there was added a top-heavy overlay of government agencies, after the fall there was an explosion of activity on the part of individuals, scientific societies, huge nonprofit organizations, and multinational entities on top of all the rest. This layering effect, which is graphically illustrated in figure 1 at the beginning of part 1 of this book, accounted not only for the quantum jumps in level of effort and funding but also in the intensity of engagement in the United States in our remarkable story of US-FSU science cooperation.

The philosophies behind the programs also changed markedly from one period to another. In the 1950s and 1960s, it was Eisenhower's vision of private citizens leapfrogging over governments, the better to understand each other and their societies. In the age of détente of the 1970s and 1980s, the Nixon-Kissinger doctrine of using scientific cooperation to influence Soviet global behavior underpinned the thinking of senior US government officials. Finally, after 1991, the predominant paradigm was that of assistance, both to prevent the dangerous proliferation of weapons of mass destruction and to sow the seeds of freedom and democracy in former Soviet lands. From thirty thousand feet, all these programs seemed similar insofar as they shared the goal of making the world a safer place. At ground level, though, the look and feel of these activities were quite different.

In the following section, I will use a very broad brush to paint pictures of each of the major US programs of scientific cooperation with (and assistance to) Russia. Each employed wholly new approaches to apply major funding to the new situation. I take a little time to describe them because to my knowledge, they have never been discussed in the same printed space and in some cases, such as CRDF, a public history of any sort has never appeared.

Major New Programs Come—and Go

The International Science Foundation

Shortly before Thanksgiving 1992, I was summoned from the NSF to a meeting at the American Association for the Advancement of Science to listen to a financier and philanthropist I had never heard of before named George Soros. He described his vision in characteristically bold and visionary terms: to provide massive, emergency support to the very best scientists in the FSU.

As is my habit, I gracelessly interrupted him, asking skeptically how much money he intended to devote to it. "A hundred million dollars," he answered, and I shut up. On December 9, he formally announced the establishment of his new project, the International Science Foundation (ISF), at a press conference in Moscow. Later that month, in the august setting of the Great Hall of the National Academy of Sciences, Soros approached me and proposed that I serve as chief operating officer for his new foundation on leave of absence from NSF. Dumbfoundedly, I muttered something like, "Um . . . yes. It would be an honor." Thus began one of the most fascinating chapters of my own career.

Many would be surprised, however, to know that coming to the rescue of Russian science was not Soros's primary goal at all. As he explained in a remarkable article in the *New York Review of Books* in April 2000,[27] his primary motivation was much more far-reaching. Science, in fact, was more a target of opportunity than a principled end in itself. At the time, Soros was deeply concerned about the appearance of what he aptly called "robber capitalism" in Russia. He had tried, but failed, to persuade the International Monetary Fund to earmark its major loan of $15 billion to make pension payments owed by the bankrupt Russian government to millions of retired Russian workers that at the time amounted to monthly installments of eight dollars. "My proposal," he wrote in 2000, "was not given serious consideration because it did not fit into the International Monetary Fund's mode of operations. So I set out to show that foreign aid could be made to work"[28]— by setting up the ISF. "My reasons for supporting scientists were complex," he continued. "I wanted to demonstrate that foreign aid could be successful, and I selected science as the field of demonstration because I could count on the support of the members of the international scientific community, who were willing to donate their time and energy for evaluating the research projects. But the mechanics of the emergency aid distribution could have been made to work for pensioners as well as scientists."[29]

There were other reasons, too, Soros acknowledged, closer to the kinds of arguments that scientists themselves would make for coming to the aid of their own colleagues, such as traditions of intellectual excellence, independent and dissident thought, and avoiding the dangers of nuclear war.[30] But while not exactly afterthoughts, they also did not occupy the center of center of Soros's goals with regard to his creation of the ISF.

Another unspoken, as far as I know, goal of Soros's in taking the bold move to set up the ISF was, I believe, impatience—impatience with world governments for their lip service to providing emergency assistance to the scientists of the FSU, and their inability to act. He did not, after all, conjure up the figure of $100 million out of thin air. It had already been floating around for a year, initially launched by the Rubbia appeal, and while Soros personally might not have known about that document, some scientists in his entourage surely did. Also, in early 1992, US congressman George Brown, launched his legislation for an "Amerus" foundation[31] that was to be generously funded at a level of $200 million over four years. But neither of these initiatives, either in the French or US governments, yielded any tangible results in a situation that was extremely urgent and time sensitive.

In this context, Soros's creation of the ISF was a classic illustration of one of the most important roles of charitable foundations, namely, to step in to address a pressing social problem when governments cannot or will not do so. Soros had done this in South Africa in the 1980s, when his Open Society Institute (OSI) was among the leading funders of the anti-apartheid campaign. With the fall of communism, recognizing the urgency of creating and strengthening civil society as a key pillar of democracy in these countries, he quickly engaged OSI in this task throughout the FSU and eastern Europe. The ISF was another replication of this pattern, but this time in the area of science, which was historically new for him and with which he often seemed uncomfortable; I heard him say more than once that funding science is the business of governments, not charities. Indeed, once the funding crisis of FSU science (particularly in Russia) had bottomed out—in large part due to his own efforts—he terminated most of the ISF's major programs and turned back to his real passion: support of civil society.

Probably the best known of the ISF's programs was its innovative Emergency Grants Program. Alex Goldfarb, who managed this program, designed a simple, quick, science-based metric to award these grants quickly and for immediate impact: any former Soviet scientist still living in the FSU who could show at least three scientific publications in the international, peer-reviewed literature received a grant of $500. Between 1993 and 1995, the ISF made 26,145 such awards, which, together with related grants recognizing collective accomplishment, came to over $15 million. These quick grants, extremely modest by Western standards but with enormous economic leverage in the context of the FSU's economic collapse, could support an entire family for a year, at a time when most scientists were receiving monthly salaries of $20 or less, or nothing at all, from their institutes.

The Soros emergency grants often made all the difference between a decision to stay in science or to abandon it, and to stay in the country or to abandon it by emigrating. The largest loss to science during this period was what Dezhina has called "internal emigration" to occupations other than science, and not "external emigration" to other countries, which was available only for that minority of scientists who were fortunate enough to have already established connections abroad.[32] For a far larger number of scientists, however, the only option to avoid starvation was internal emigration—leaving their institutes to work as at menial but paying jobs. And in the fast-moving world of science, even a temporary absence meant that one would become hopelessly behind on the literature, on any pretense of publishing, and very often to their profession as scientists.

What distinguished the Soros grants, above all, was the strict and highly innovative adherence to standards and measures of scientific quality. This was absolutely revolutionary in a national science system that awarded research support through a top-down, command-economy system. Each major Soros program, moreover, applied different approaches in a much more diverse overall portfolio than any Western research grant program of which I am aware. The Emergency Grants Program applied the rigid, quantitative measure of scientometrics to make quick, small awards to deserving scientists. The International Soros Science Education Program, carried out independently by Valery Soyfer, focused on outstanding Russian university professors and high school teachers and offered smaller prizes for students, sponsorship of "Soros Olympiads" engaging over 800,000 students, and a monthly educational journal. Multiyear grants for professors and teachers, ranging from $250 to $500 per month, were awarded based on survey evaluations of former university professors' students about the quality of their teachers' instruction.

The Long-Term Grants Program, the largest of the ISF programs with over $80 million of support, applied the technique of competitive merit review, also known as peer review, to the selection of investigator-initiated, small-team research grants. Aside from the extremely small programs of the Russian Foundation for Basic Research, which had been started under the leadership of mathematician Andrey Gonchar in late 1992, the Long-Term Grants Program marked the first time that peer review was used as a major funding technique in the Soviet Union,[33] and certainly the first time with such an enormous amount of money. In chapter 7 I will return to this important theme when we discuss one of the most intangible results of international scientific cooperation: institutional and cultural change.[34]

The ISF Long-Term Research Grants Program, which was my personal responsibility, started in late 1995. Over the course of two competitions in eighteen months, it received over thirty-five thousand proposals, which it mailed to normally no fewer than four individual reviewers per proposal, mainly in the United States, and then submitted to sixteen international scientific review panels for their recommendations. It awarded 3,555 competitive, merit-reviewed awards ranging in size from $27,000 to just under $10,000 to small FSU research teams, resulting in nearly $80 million in awards, the largest single ISF activity.

Table 4.1. ISF Programs[i]

ISF Program	Amount ($millions)
emergency grants	$15.585
long-term grants	$80.072
conference travel grants	$14.367
library assistance	$4.777
telecommunications development	$4.141
TOTAL	$126.787

i. Dezhina 2000, 151.

To put these statistics in context, at the time, the number of proposals received by the ISF for this competition was about equal to the total number of proposals received by the NSF in an entire year. Individual experts serving on review panels contributed 22.5 person-*years*[35] of work to rank and recommend long-term research grant proposals for award, and this does not count what I would estimate as the work of some fifty thousand "mail reviewers" of these proposals. Scientific societies (such as the APS, AAS, AMS, American Society for Microbiology, American Chemical Society, and NAS) provided invaluable support in suggesting mail reviewers and appointing the panels that recommended the final distribution of awards.

Finally, within the ISF there was also a conference travel program to enable former Soviet scientists to attend international scientific meetings; a library assistance program that distributed physical copies of Western scientific journals to science libraries throughout the FSU; and a telecommunications program that established many of the first internet linkages for former Soviet civilian scientific institutes and provided them with basic equipment and cable to establish computer networks.[36] Table 4.1 gives an overall breakdown of these programs.

In addition to these funded ISF programs, in 1994 the ISF initiated a facilitative program that took advantage of the Russian tax exemption for foreign research grants to make available its secure and transparent financial management system for transferring funds to FSU recipients from other nonprofit organizations for research purposes. Over its lifetime, the ISF's Grant Assistance Program, or GAP, handled transfers of $30 million in external funds to FSU recipients. Initially, the program performed these services at no charge, and in 1995 Soros even added his own $3 million contribution to underwrite the administrative expenses of managing these funds. After 1995, when the program was briefly transferred to the Open Society Institute, the GAP levied a 10 percent administrative fee on most donor clients but waived it for smaller ones.[37]

The GAP was a major administrative innovation that made it possible for a broad range of third parties to take advantage of a highly sophisticated and scrupulously managed system for getting money and equipment to scientists in countries where banking systems were primitive and about which there were widespread and often well-justified concerns about corruption. Once CRDF was established, it first used ISF's GAP system to transfer its own funds and, ultimately, redesigned and launched its own system, which performed similar services and became a significant source of additional revenue for CRDF in its own right.

The Nonproliferation Programs

Until 1995 the ISF was the major US source of serious funding for civilian, nondefense science during these years of financial crisis. In the meantime, however, important new programs with major US government funding, both multilateral and bilateral, emerged to address the worrisome impact of Soviet science's collapse on global security and the potential proliferation of weapons of mass destruction—their physical materials, their technologies, and the scientists who created them—to rogue states and other bad international actors.

The "Science Centers": ISTC and STCU

Apart from the carryover of the intergovernmental agreements to the Russian Federation under the Gore-Chernomyrdin Commission, the first major new government-sponsored effort to get underway in this period was the International Science and Technology Center, or ISTC.[38] The ISTC, while it may have been initiated by the US government, was not a bilateral project; it was a multinational organization, similar in diplomatic standing to the United Nations. The original signatories of its founding document were the European Union, the United States, Japan, and the Russian Federation. The ISTC also spawned a sister organization in Ukraine, the Science and Technology Center Ukraine, or STCU; while independent of ISTC, it shared its mission and received its US share of financial support through the same congressional authorization and appropriation process.

Nearly the entire territory of the FSU was covered by the complementary geographic jurisdictions of the ISTC and STCU. The ISTC was responsible for activity in Russia, Kazakhstan, Armenia, Kyrgyzstan, Tajikistan, and at times Georgia, and the STCU was responsible for work in Ukraine, Uzbekistan, Georgia, Azerbaijan, and Moldova.

Together, the ISTC and the STCU were known informally, and especially in the government community, as the "Science Centers." In a sense this was a misnomer because while they certainly did support scientific research and scientists, they were certainly not the only "science" show in town. Yet for the government community in the broadest sense—Congress, the defense community, and the diplomatic service—the Science Centers represented, in fact, the US government's primary commitment to the support of former Soviet scientists after the fall of the Soviet Union.

ISTC's primary objective was to give former Soviet WMD and missile scientists the opportunity to redirect their work toward peaceful activity. Its principal approach was to provide very large research grants on the order of $400,000 a year to teams of former WMD scientists to work on nonmilitary research. Schweitzer reports that in 1994, its first operational year, such projects totaled $50 million and that from 1994 to 2000, funding of ISTC research projects exceeded $314 million. This running start was followed, in Schweitzer's words, by an "era of euphoria" from 2001 to 2006, in which project funding averaged between $68 and $78 million dollars annually. The core research grant program was augmented by seminars, support for conference travel, and more targeted programs such as those aimed at teaching business innovation skills.[39] There was also a "partner program" that brought in participation—in cash, kind, and personal—from third-party foreign sources such as corporations, universities, and nonprofits interested in supporting ISTC's grand mission and taking advantage of the scientific and technical capabilities of those whose work and redirection it funded.[40]

ISTC's funding was reasonably well distributed among the Western partners. Between 1994 and 2011, of the roughly $859 million spent by the ISTC, $243 million (28%) came from the EU, $223 million (26%) from the US government, $64 million (7%) from Japan, and another $58 million from the governments of Canada, the Republic of Korea, Sweden, Norway, Finland, and others countries. Impressively, the largest single funding source was not a country but the *category* of "partners," which contributed nearly $271 million, or 32 percent, of the ISTC's expenditures over its lifetime.[41]

Glenn Schweitzer, the ISTC's first executive director and its historian, counts the ISTC's waning years from 2007. A variety of factors contributed to its decline. Briefly, they were political (Vladimir Putin's increased assertiveness in international relations and waning enthusiasm for assistance to Russia in the US Congress), financial (budget squeezes in the United States motivated the State Department to begin "graduating" institutes from the program to save money), and what I would call cultural (resentment in Russia to being perpetual objects of unilateral assistance as opposed to peers in cooperation). These factors, and others, affected all the major post-1991 programs to one degree or another, to the extent that they survived long enough. In 2011, the Russian Foreign Ministry formally notified the ISTC of Russia's intention to withdraw its membership in 2015, and in that year the ISTC relocated to Kazakhstan and continued to operate, though without Russian membership.

In Ukraine, the STCU continued to operate uninterrupted, but with time, it became clear that demography tended to vitiate its mission. To the extent that the mission of the STCU (as well as that of the ISTC) was to redirect or "engage" (the latter became the term of preference in policy circles) former WMD scientists, the eligible pool of recipients dwindled to zero with the aging of that population. As a result, direct US funding of the STCU came to an end in 2015, although it continued to serve as a conduit for flow-through "partner projects."

Global Initiatives for Proliferation Prevention

The informal relationships that came to be called the "lab-to-lab program" did not have an official basis in legislation until the program became formalized in 1994 in a new program with a budget line item approved by Congress. Perhaps the most unusual feature of this legislation—the Foreign Operations, Export Financing, and Related Programs Appropriation Act—was the provision to allow agreements with private US companies, to include cost-sharing, as part of the program's implementation.[42]

The US Department of Energy (DOE), which received the appropriation as the parent of the national labs referred to in the legislation, dubbed the new program the "Industrial Partners Program" and designed a highly innovative approach that firmly integrated the profit motive into a cooperative science program with the FSU in a manner in which no other program had done before or since. Each project was to involve a team of former Soviet WMD scientists or engineers, a scientist or team from one of the DOE national laboratories (e.g., Los Alamos, Livermore, Oak Ridge, Brookhaven), and a for-profit US high-tech company.

The DOE funding was allocated between the former Soviet lab and the US national lab, while the private company was required to pledge a 100 percent cost-share—in a mix of cash and kind—to the government's contribution. Each project's goals would be to develop a civilian technology that could be developed for commercial use, generating revenue for both parties, and, especially, to secure civilian employment or at least income for the FSU participants. No funds passed from the government to the US partner company. Instead, the incentive for the companies was to gain access to the FSU scientists' technical capabilities, in the knowledge that the latter would be getting reliable financial support from the DOE for the project's duration. In 1996, the DOE changed the name of the Industry Partnership Program (IPP) to Initiatives for Proliferation Prevention (also IPP), and later on tacked on the word "Global" at the beginning to round out the acronym to GIPP, which remained the program's title until its demise in 2015. To interface with the private US companies, in 1994 the DOE also created and funded a new nonprofit organization, the US Industry Coalition (USIC), which was a membership association of the participating US companies.

Over its lifetime, the GIPP program made grants of over $250 million to former Soviet WMD scientists in the so-called closed cities and other institutes involved in weapons of mass destruction research. Roughly equal cost shares were made by the partner US companies, though this was difficult to track because many of the cost shares were largely in kind (time and effort of their own personnel, gifts of equipment to their FSU partners, and the like), not cash. According to data carefully compiled by USIC over the years in cooperation with the DOE, GIPP projects with US industry partners had a commercialization success rate—measured by sales, contract work, and other revenue—in excess of 25 percent, and the US companies were able to attract follow-on investment funding for the technologies developed under the program of more than $250 million.[43]

Appealing to the profit motive to advance nonprofit or public purposes is a difficult art, but the unique structure of the GIPP program was actually quite successful in making that happen. In fact, when I joined USIC in 2006, I was quite skeptical at first of its claims of commercial success through the GIPP program. But once I understood firsthand the rigorous standards USIC had developed to gauge companies' actual commercial achievements, I became convinced that this relatively small program (the total GIPP budget never exceeded $50 million a year and was usually at most half that amount) had indeed found a formula for significant leverage in technological innovation. This was all the more impressive since this all happened in the context of the most difficult-to-engage scientists from the most heavily guarded institutes of a highly sensitive field in a region in which rule of law and market competition were embryonic at best.

The GIPP program's success, and its greatest vulnerability, were both products of its mixed goals and its unusual complexity. Its primary mission was nonproliferation of WMD technologies and creating incentives for the scientists and engineers who developed those technologies from going to or working for rogue states and other hostile actors. But with the explicit goal of achieving tangible commercial results, its metrics became mixed and confusing. One of the key vulnerabilities of all the nonproliferation programs was the logical impossibility of "proving a negative." While it was virtually impossible to demonstrate that any given number of WMD scientists were *deterred* from offering their services to hostile states or groups, showing the number of actual new civilian research-related jobs created in the FSU and even the United States had at least some positive value. However, despite the impressive statistics on funds raised by the US industry partners, the number of verifiable new jobs created through the program was actually rather modest.

Additionally, for the DOE labs, and especially for the DOE's more traditional government managers and officials, this must have been an uncomfortable and somewhat artificial marriage. Not only did the GIPP program challenge the customary lines of demarcation between government and industry, but it also, by introducing a commercial metric for the success of a nonproliferation program, had the unintended effect of blurring the program's core security goals to such an extent that it became difficult for Congress, which commissioned numerous investigations of the program by the Government Accountability Office, to fairly evaluate the program's outcomes.

The combination of security and commercial goals was a winning formula so long as the relationship with Russia, in particular, was positive and so long as there continued to be a perception of a strong threat to global security from the FSU WMD community. With the passage of time, however, the perception of threat decreased and the program's very complexity became its undoing. The DOE struggled valiantly but in vain to explain the program and its achievements to skeptical congressional committees. It was finally terminated in fiscal year 2014.

The Civilian Research and Development Foundation

Congressman George E. Brown Jr., a Democrat from southern California who served in the US House of Representatives for thirty-five years, was without question one of the boldest and most imaginative advocates of science in the history of the US Congress. A longtime member of the House Science and Technology (S&T) Committee and its chairman between 1991 and 1995, Brown left a legacy of creative initiatives that reflected his passion for science, public policy, and international affairs. Domestically, his accomplishments included the legislation creating the Environmental Protection Agency, the Office of Science and Technology Policy (also known as the "White House Science Office"), and the Office of Technology Assessment. In the international arena, as House S&T Committee chair, he focused his imagination and legislative skill on two unprecedented initiatives to create "foundations" to promote international scientific cooperation—one with Mexico and one with the FSU, the latter of which became what is now known as CRDF Global.

CRDF's founding legislation, after one false start, came in an amendment offered by Rep. Brown to the 1992 FREEDOM Support Act,[44] which laid the official basis for the entire US assistance effort to the FSU and which enjoyed broad bipartisan support. The goals for the new foundation reflected a broad hybrid of policies embraced by the US government with regard to the FSU at the time:

(1) To provide productive research and development opportunities within the independent states of the former Soviet Union that offer scientists and engineers alternatives to emigration and help prevent the dissolution of the technological infrastructure of the independent states.
(2) To advance defense conversion by funding civilian collaborative research and development projects between scientists and engineers in the United States and in the independent states of the former Soviet Union.
(3) To assist in the establishment of a market economy in the independent states of the former Soviet Union by promoting, identifying, and partially funding joint research, development, and demonstration ventures between United States businesses and scientists, engineers, and entrepreneurs in those independent states.
(4) To provide a mechanism for scientists, engineers, and entrepreneurs in the independent states of the former Soviet Union to develop an understanding of commercial business practices by establishing linkages to United States scientists, engineers, and businesses.

(5) To provide access for United States businesses to sophisticated new technologies, talented researchers, and potential new markets within the independent states of the former Soviet Union.[45]

The responsibility for establishing the new foundation was assigned to the director of NSF, who was also authorized to receive funds from other sources for its operation. On August 11, 1995, NSF director Neal Lane formally established CRDF, appointing a board of directors for the nongovernmental foundation, with NAS member and highly respected advocate of international science Peter Raven as chairman. He also asked me, having just returned from detail to the ISF, to be the foundation's first executive director. And so began another highly interesting chapter in my checkered career.

CRDF began its existence with a modest budget of $10 million, which came in equal parts from the NSF and the US Department of Defense.[46] The NSF contribution derived from an undesignated gift from none other than George Soros, who was persuaded by Vice President Al Gore to make the bequest in preparation for a summit meeting between President Bill Clinton and President Boris Yel'tsin of Russia.[47] With such modest resources, obviously unsuitable as an endowment corpus, CRDF decided to spend it all in a short period of time to make an impact and then see what would happen.

Many of CRDF's first programs were in fact a logical extension of proven concepts developed and implemented in the ISF, most notably a competitive cooperative grants program, but this time not for direct assistance to FSU scientists for their own work, but now for bilateral civilian joint research projects between the United States and the FSU. CRDF also launched an innovative effort to address its mission's business R&D mandate, and it assisted the State Department under contract in arranging visits to the United States by former Soviet WMD scientists. By 1997, we had virtually exhausted our funding, and CRDF senior vice president Charles T. ("Tom") Owens, a former NSF colleague, and I were poised to hand out pink slips to our small staff of about fifteen and close up shop.

Two events then converged to breathe new life into CRDF, in addition to the continuation of our contract work for the State Department mentioned earlier. First, the State Department's Cooperative Threat Reduction Program, which had been assigned responsibility for technical activities under the FREEDOM Support Act, asked CRDF to manage a small program to create an international geophysical research center in Kyrgyzstan in fulfillment of a pledge made by Vice President Gore to Kyrgyzstan President Oskar Akayev. The Kyrgyzstan project was a success and, together with the nonproliferation support that CRDF was providing for the State Department, generated sufficient confidence in CRDF's capabilities to justify honorable mention in congressional committee reports that in turn encouraged the State Department to allocate a regular annual subvention to CRDF, which grew fairly quickly to the level of $15 million a year. This FREEDOM Support Act funding became CRDF's core source of support for many years, underwriting its cooperative grant program as well as its Next Steps to the Market and other competitive activities with the countries of the FSU.

The second life-saving event for CRDF was a series of telephone calls from Victor Rabinowitch, senior vice president of the John D. and Catherine T. MacArthur Foundation, to both Loren Graham at MIT and to me at CRDF asking what we thought might be the next important thing to do with Russian science now that the "crisis" was over. With a modest planning grant from MacArthur, CRDF convened a panel of experts from the United States and Russia, including Graham, Harley Balzer of Georgetown University, Glenn Schweitzer of the NAS, Irina Dezhina of the Institute for the Economy in Transition in Moscow, and Kristin Wildermann and myself from CRDF staff to study the issue and generate a report. We later added Russian Federation minister of science and technology policy Boris Saltykov and Russian Academy of Sciences member Mikhail Alfimov to ensure that the idea represented both American and Russian understanding of real needs.

The report, "Basic Research and Higher Education in Russia [BRHE]: A Proposal for Reform," proposed a $50 million five-year program of major grants to Russian universities to establish interdisciplinary Research and Education Centers (RECs) whose goal would be to integrate research and teaching at Russian universities, reversing the highly vertical organization of Soviet research into elite academy institutes and bifurcating that research from university teaching. It also recommended a $10 million five-year program of generous Exceptional Young Investigator Grants to provide incentives and support for outstanding young researchers at the graduate level to pursue their studies at Russian universities and, it was to be hoped, embark on brilliant careers as university teaching and research professors.

Sixty million dollars, however, was a very tall order, and the MacArthur Foundation sensibly stipulated that the cost would have to be shared by the Russian side. In 1997, we thought the chance of that happening was infinitesimal to zero. But then something quite amazing happened. In late 1997, Rabinowitch and I paid a courtesy visit on Aleksandr Tikhonov, the Russian minister of education, in Moscow. We explained the BRHE concept to Tikhonov but lamented the prospect of getting it funded without matching funds from Russia. Then, to our great surprise, Tikhonov quietly said, "We'll fund it." We were floored. If ever an international science program received confirmation of a partner government's commitment, this was it. We didn't quite believe it, but we came back home and CRDF wrote a proposal jointly addressed to the MacArthur Foundation and to the Carnegie Corporation of New York, which was supporting a somewhat different project to support social science research at Russian universities.

With the collapse of the ruble in August 1998, touching off a major economic crisis in Russia, we were sure that the project was dead. But the Russian Federation's Ministry of Education and Sciences (commonly known as Minobrnauka), now led by Andrey Fursenko, a reform-minded science organizer from St. Petersburg whom some of us knew personally,[48] kept signaling us that it was still on board. And it stayed on board. In particular, a young Russian nuclear physicist, Mikhail Strikhanov,[49] who was both passionate about the program and an excellent organizer, managed year after year to come up with the matching funds.

In the end, the BRHE program competitively selected twenty RECs at universities from all across Russia, each with a link to one or more institutes of the Russian Academy of Sciences and each with an average initial three-year grant of $1 million to $1.5 million and often two-year extensions as well. For the initial sixteen centers, about $45 million was disbursed, with more than half coming from the US side, one-quarter from the Russian federal government, and one-quarter from local Russian sources. The final four centers were funded entirely by the Russian side.[50]

What is more important, the Russian government, which had become strongly committed to dismembering the powerful, independent Russian Academy of Sciences for reasons of its own, embraced the basic principle advocated and established through the BRHE program—the integration of research and education—by pivoting its budget support from the academy to the universities, generating a host of programs to strengthen university research and teaching and to create "national research universities" in Russia that would be able to compete on a global scale with the great Western institutions of higher learning. The BRHE program was finally phased out in 2013.

The BRHE program was probably the most innovative, and certainly the most ambitious, program that CRDF conducted in any country to that time. From my vantage point, it should probably also be counted as one of the more successful international programs ever undertaken to effect institutional change in the science system of another country. The key reason for this success was that it coincided with, and reinforced, the Russian government's own decision to rearrange its research establishment, shifting the emphasis from the academy institutes to the universities.

Additionally, unlike most of the other major programs in this period, it was jointly governed through a bilateral governing board and jointly funded. Its governing board, in turn, appointed a high-powered US-Russian Expert Committee, chaired by Aleksandr Dykhne (to whose memory this book is dedicated),[51] to evaluate and certify the major grants to the Research and Education Centers. Later on, an independent, international evaluation panel convened by CRDF in 2007 called it "the right program" at the "right time" with the "right process."[52] CRDF also replicated the concept on smaller scales in Georgia, Armenia, and eventually Ukraine with targeted support for selected higher educational institutions, albeit without the joint governance feature of BRHE. In addition to these university-based program, CRDF also launched other "institution-building" projects, such as creating local CRDFs in several countries: Georgia, Armenia, Azerbaijan, and Moldova.

CRDF's role, however, was not solely focused on science in the civilian sector of the FSU. Second in its list of goals, cited earlier, was "to advance defense conversion by funding civilian collaborative research and development projects between scientists and engineers in the United States and in the independent states of the former Soviet Union." CRDF did this both through its core cooperative research programs as well as through contract services for the Departments of State, Health and Human Services, and Defense in support of their own cooperative threat reduction programs.

CRDF also took over the legacy of the Grant Assistance Program (GAP) from George Soros's Open Society Institute and subsequently developed its own extensive service program, which it branded as "CRDF Solutions." With time, CRDF also broadened its geographic focus, for example taking on major responsibilities for implementing the scientific and technical components of President Barack Obama's Cairo Initiative with Arab countries of the Middle East. At that time, it also rebranded itself as CRDF Global, signifying that it was moving beyond its initial mandate to work

Table 4.2. CRDF Global Funding of US-FSU Scientific Cooperation, 1996 to July 2016[i]

Country	Collaborative Research	Travel and Conference Support; Fellowships	Capacity Building and Higher Education	Entrepreneurship and Innovation	TOTAL
Armenia	$3,510,249	$1,381,307	$3,400,574	$1,070,602	$9,362,733
Azerbaijan	$1,029,078	$53,044	$1,993,628	$306,586	$3,382,336
Belarus	$474,686	$18,408			$493,094
Georgia	$4,713,409	$902,117	$3,424,958	$1,260,318	$10,300,802
Kazakhstan	$1,256,490	$204,440	$1,466,406	$281,056	$3,208,392
Kyrgyzstan	$563,561	$93,024	$1,363,936	$57,715	$2,078,235
Moldova	$3,012,534	$797,214	$2,693,203	$368,965	$6,871,916
Russia	$31,856,057	$1,667,504	$24,029,352	$7,623,974	$65,176,888
Tajikistan	$99,455	$47,898	$232,090	$47,939	$427,383
Turkmenistan	$52,635		$28,041	$23,664	$104,340
Ukraine	$8,643,370	$956,534	$1,316,867	$3,028,758	$13,945,529
Uzbekistan	$1,464,866	$284,483	$844,675	$206,538	$2,800,562
TOTAL	$56,676,391	$6,405,974	$40,793,729	$14,276,115	$118,152,209

i. Source: CRDF Global, personal message, dated August 25, 2016.

with the countries of the FSU and extending its effective approaches to other regions.

Table 4.2 summarizes CRDF Global's total spending on cooperative activities with the countries of the FSU from inception in 1996 through July 2016, for a total of over $118 million. The funds shown in the table are only those disbursed to the FSU side, including individual financial support, purchase of equipment and supplies, and other costs. In addition, the FSU country partners contributed some $39 million in matching funds, the preponderance of them (nearly $25 million) from Russian federal and local sources in connection with the BRHE program. CRDF Global also provided support for US participant travel and other direct expenses, but not salary, at a cumulative level of more than $9.8 million during the same period, which was supplemented by another $2.5 million in third-party funds from other US and non-FSU sources. In recent years, funding for activities for the region in general, and for Russia in particular, have declined as the organization's focus has shifted to other world regions, especially the Middle East and North Africa, and as a result of declining US government support for cooperative activities with Russia since 2014.

Where Do Things Stand Now?

Since the Russian annexation of Crimea and invasion of Ukraine in early 2014, most official US government–managed and funded cooperative science programs with Russia have either slowed substantially or ground to a halt.

The extent of this decline can be seen from an analysis of "Russia-related" grant awards made by the NSF in figure 4.1. This data, drawn from the publicly available NSF awards database, shows the total number of grants made by the NSF to projects with any identifiable link to Russia—whether in the title, abstract, or other required programmatic or geographical coding.

It is important to keep in mind that such data are incomplete because they do not report the informal links and cooperation among scientists that take place independently of given funding instruments. There is no outright ban on scientific contacts between US and Russian scientists, nor is there ever likely to be, and where personal relationships have already been established, they very well may continue. But at the same time, the trends in government-funded research projects (in the basic sciences, in this case) reflected by these data are unmistakable.

Another important and discouraging development in the recent US-Russia standoff was the closure of CRDF Global's Moscow office in early 2017. This occurred in part because of long-standing Russian government persecution of foreign-funded nonprofit organizations under Vladimir Putin and in part due to the overall decline in program activity. This step marked the end to a more than twenty-two-year presence, during which CRDF Global not only supported direct grants for cooperative research but also provided substantial grant management assistance to many other organizations, both governmental and private, including NSF. With the disappearance of this widely used facilitative presence in Moscow, it is likely that a great deal of cooperative activity funded by other organizations will cease as well. Additionally, since the CRDF Global presence in Moscow was important in facilitating field research in the geosciences, which make up over half of NSF-funded activities involving Russia in general, the ramifications of the Moscow office closure could have a significant multiplying effect.

Former Los Alamos director Sig Hecker, in his epic retrospective of US-Russia cooperation on nuclear nonproliferation, wrote the following about the decline and virtual end of collaboration between US and Russian weapons

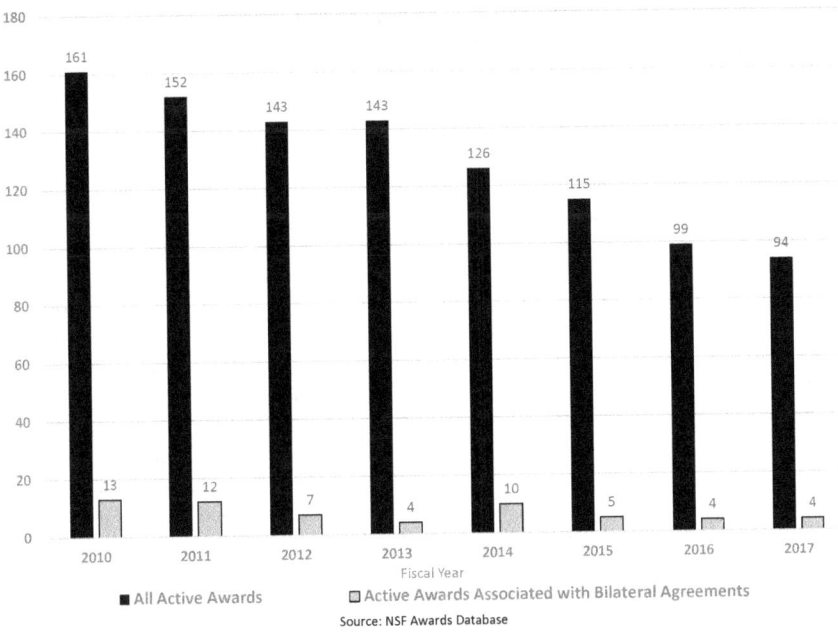

Figure 4.1. Number of Active NSF Awards Involving Russia, FY2010–FY2017

scientists, in which he and many others had invested such enormous efforts after the Soviet Union's collapse in 1991: "Scientific collaborations flourished into the first decade of this century. However, they declined as the relationship between Moscow and Washington became more strained and both governments made visits and access to facilities more difficult. Unfortunately, the relationship was strained to the point of crisis in 2014, resulting in almost all collaborations coming to an end, much to the dismay of those of us who have nurtured the relationships for more than two decades."[53]

With this, we come to the end of this narrative overview of sixty years of formal programs of scientific cooperation between the United States and the Soviet Union, including the Russian Federation during the post-Soviet period. In these few pages it has hardly been possible to do justice to any one of these programs, and I have no doubt that there are other worthy efforts that I have omitted entirely. As I said at the outset, my purpose in this chapter was not to be exhaustive or to capture every detail of these programs but instead to give the reader a sense of their nature, rationale, scope, and dimensions in order to set the stage for the testimony of my interview participants in part 2 of this book.

Notes

1. "Putin Deplores Collapse of USSR," *BBC News*, April 25, 2005.
2. The best overall treatment of the USSR dissolution's impact on scientific research is Graham and Dezhina 2008, and I draw liberally from it throughout this section. *Science* and *Nature* magazines, particularly the former, had ongoing news coverage and occasional in-depth interviews on the subject. As appropriate, I will cite these and other sources during the discussion. As I was deeply involved in US efforts to address this crisis, events not otherwise cited in this section are from my personal notes and recollections.
3. Estimates of the number of scientists and engineers in the Soviet Union varied considerably, depending in large part on how one defined the category of "engineers." NSF's Science and Engineering Indicators for years gave high and low estimates, which, in any case, exceeded the size of any other country's scientific community. NSF's *Science and Engineering Indicators 1987* (SEI 1987), the last issue in which this data was tracked, reports that as of 1985, the "low estimate" for the USSR was 1.5 million scientists and engineers engaged in research and development (or 96.9 per 10,000 of the total labor force population). SEI 1987 shows the Soviet 1965 high estimate as 1.7 million, or 112 per 10,000. For comparison, the next largest scientific and engineering community in the same year was that of the United States, standing at 790,000 individuals and 65 per 10,000 of the active R&D labor force population, or conservatively just more than half that of the Soviet Union. See appendix tables 3-17 and 3-19 in *Science and Engineering Indicators 1987*, 227–228.
4. Graham and Dezhina 2008, 1.
5. In this respect, however, the Soviet Union's scientists were different only in degree to scientists the world over, who are often the worst marketers of their own results. The difficulties of even some of the best scientific laboratories in the West, such as the DOE's national laboratory system, have been legendary. On the chronic difficulties faced by Russian science from its very beginnings, see Graham 2013.
6. Cohen 2000.
7. This agreement had initially been signed with the Soviet Union, and the Russian Federation took over responsibility for it after the USSR's dissolution.
8. Interview with John Zimmerman, September 30, 2015.
9. These words, though they did not appear in the actual letter from Rubbia to Mitterand, were cited by Dezhina (2000) as the title of the appeal; as I recall, this was indeed the actual title of the initial document that was sent by Okun' and his group to Rubbia.
10. Carlo Rubbia to François Mitterand, letter dated September 26, 1991. I am grateful to CERN for its permission to quote short passages of this letter.
11. Dezhina 2000, 8.
12. Rubbia to Mitterand, 1. The translation is my own.
13. Ibid.
14. Dezhina 2000, 8. In personal conversations, Irina Dezhina has told me that the source of the text citing the $100 million figure was Lev Okun' himself. I had written to Okun',

with whom I had a personal relationship, in late 2015, to seek verification. Sadly, however, he passed away from a long illness on November 23, 2015.

15. Cooperation was intended for the other countries of the former Soviet Union, as well. But because the center of gravity of Soviet science was in Russia (and note that this was the focus of the Rubbia-Okun' appeal), there was an informal tendency to gravitate attention toward that country. Most US and international efforts, though, were careful—and properly so for scientific and other reasons—to devote corresponding proportionate attention and resources to the science communities of the other former Soviet republics.

16. *Reorientation of the Research Potential of the Former Soviet Union: A Report to the Assistant to the President for Science and Technology* 1992, 1.

17. Ibid., 1–4.

18. In this context, it is also important to understand the difference between the emergency assistance provided early on by the scientific societies and the Soros program, for example, and the subsequent US government–funded programs. The former were deliberately of a time-limited nature; the latter, however, were conceived as long-term efforts, and as I will argue later, they continued far too long after the very concept of "assistance" was appropriate.

19. Siegfried Hecker's new two-volume collection of essays and interviews with Russian and American scientists who worked together in the lab-to-lab program is an immense contribution to documenting and understand this critically important program. See Hecker 2016.

20. Interview with Siegfried Hecker, January 28, 2016.

21. The following discussion is drawn largely from Dezhina 2000, 12–18.

22. Ibid.

23. Ibid., 10

24. AmeRus Foundation for Research and Development Act of 1992.

25. Dezhina 2000, 10.

26. This was the acronym created by INTAS itself, from the words "international association."

27. Soros 2000.

28. Ibid., 6.

29. Ibid.

30. Ibid., 6–7.

31. AmeRus Foundation for Research and Development Act of 1992.

32. Dezhina 2002.

33. Some Academy of Sciences research institutes did have internal research grant competitions; as I recall from a conversation with Ioffe Physico-Technical Institute (St. Petersburg) Zhores Alferov, that institute did so. But these were exceptions based on the institute director's personality and the nature of the review process probably differed significantly from conventional Western practices.

34. See chapter 7, "Impact on Scientific Infrastructure."

35. The American Psychological Association defines a "person-year" as a unit of measurement, especially in accountancy, based on an ideal amount of work done by one person in a year consisting of a standard number of person-days. In this context, the calculation is based on an average US government work year consisting of 208 days.

36. Dezhina 2000, 64–104.

37. Ibid., 143–150.

38. Schweitzer 2013 provides an excellent and exhaustive study of the ISTC, from which I draw in this section.

39. Ibid., 27–28, 35.

40. In this sense, the "partner program" was similar to the ISF (and later CRDF) Grant Assistance Program, but with a more industrial focus.

41. It might be more accurate to characterize the "partner" contributions as flow-throughs rather than an actual component of the ISTC's budget, much as the ISF and CRDF grant assistance programs provided such facilitative services in the civilian science sphere. In any case, this number is certainly impressive.

42. Foreign Operations, Export Financing, and Related Programs Appropriation Act (1994), Pub. L. No. 103-87, Sec. 575, 107 STAT 972–773 (1993).

43. *USIC E-Notes*, vol. 12, no. 2 (September 2011).

44. FREEDOM Support Act, Pub. L. No. 102–511 106 Stat. 3320 (1992).

45. Ibid., Sec. 511(b).

46. In the FSA, the primary funding source designated in the legislation was the so-called Nunn-Lugar Program, under the National Defense Authorization Act of 1993, Title XIV, Subtitle E.

47. This extraordinary action may have been the only time that Soros made such a gift to the US government—or any other government, for that matter.

48. Fursenko was from Vladimir Putin's inner circle, as later events made abundantly clear. In March 2014, he was one of nineteen Russian individuals named by the United States as sanctioned individuals, in retaliation for the Russian annexation of Crimea. Almost all those named were Putin cronies. Subsequently, Fursenko became Putin's chief science and technology adviser and played a key role in the demise of the Russian Academy of Sciences. See chapter 10, "On Science Reform."

49. Now Rector of the Russian Federation's National Federal Research University MEPhI (Moscow Engineering-Physics Institute).

50. Graham and Dezhina 2008, 118.

51. No more conclusive evidence that the program was welcomed by the Russian government was the personal telegram from President Putin to Aleksandr Dykhne, the chair of the program's Expert Committee, displayed on the dedication page of this book.

52. "Integration of Teaching and Scientific Research in Russia: An Independent Evaluation of the Basic Research and Higher Education Program 1998–2007," August 2007.

53. Hecker 2016, 2:186–87.

PART II

IN THEIR OWN WORDS

5

HOW DID IT START?

Introduction

In this chapter, we begin looking at the firsthand testimony of the participants themselves—the scientists, program managers, government officials, and others who took part in cooperative science activities with the former Soviet Union (FSU) beginning in the late 1950s and through the first two decades of the 2000s.

In my interviews, I used a series of four open-ended questions through which I hoped to gauge the respondents' initial hopes and intentions, their actual experiences, and their retrospective evaluations of how they succeeded or failed in realizing those intentions—and what surprises they may have encountered along the way. These were:

- How did you become engaged in US-FSU science cooperation, and what were you thinking? What were the motivations that kept you engaged?
- How did it go? What were your main achievements and disappointments? What problems did you encounter?
- Looking back on it now, how do you assess your initial motivations?
- What are your favorite stories?

While the questions themselves may seem simple, the answers are as varied and sometimes as surprising as human nature itself. In this chapter, the participants share with us, in their own voices, their answers to the first question: How did you become engaged.

The Scientists

Kip Thorne

The first, and most presumably obvious, way in which scientists embarked on science cooperation with the FSU was through science itself. Nobel

Laureate Kip Thorne shared with me in great detail how and why he became involved. His story includes many strands that we will recognize in the accounts of other scientists as well.

Like many scientists, Thorne's first exposure to scientists from the Soviet Union occurred at an international scientific meeting:

> I finished my PhD in June 1965. I had worked in relativistic astrophysics, the interface of relativity and astrophysics. This was a new field, a very new field. The major centers were Princeton . . . Moscow . . . and England.
>
> Right after I finished my PhD, I went to the Fourth International Conference on General Relativity and Gravitation in London, which was in July 1965. Of course, all the eminent Soviet physicists had been invited and of course [Yakov] Zel'dovich was not allowed to come out for obvious reasons, but Igor Novikov, a collaborator of Zel'dovich, did get out; he was my age, maybe just a few years older than me. He had completed his PhD just a few years before me. And [Vitaly] Ginzburg was there, and I don't remember whether [Iosif] Shklovskii got out. But I met Novikov and we became close friends very quickly. We had admired each other's publications.[1]

Three years later, Thorne made his first visit to the Soviet Union and began having "extensive interactions" with fellow theoretical physicists in Moscow, whom he met through Novikov. Novikov also visited Caltech in 1968. As Thorne recalled, "The first time that I went to the Soviet Union was in '68, to a relativity conference in Tbilisi followed by several days in Moscow. And later that same year, I got Novikov out for an extended visit. I can identify these visits precisely because *2001: A Space Odyssey* was released in theaters in 1968 during Novikov's visit to California, and we went to see it."[2] (Such is the way astrophysicists mark the passage of time, apparently.)

Novikov's trip to the United States was unusual, however, in two senses. First, in those years, Soviet theoretical physicists did not travel internationally a great deal. This was probably the case because their work was so advanced in terms of world science as a whole, and often related to military research, that they had a harder time getting permission to travel abroad than experimental physicists or most other scientists. Another serious issue was that many theoretical physicists, as well as mathematicians, were Jewish and for that reason alone were routinely not issued exit visas.

Novikov, however, was not Jewish, and he was able to accept Thorne's invitation. But the second thing that strikes me as unusual about his visit is that at this early date it took place outside the framework of the formal exchange programs; the invitation was issued in a direct letter from Caltech. Within the National Academy of Sciences (NAS), where I worked from 1973 to 1978, it was the reigning assumption that the two countries' top scientists relied on the interacademy program to move back and forth. This was only partly true. My sense is that theoretical physicists, and to a lesser extent mathematicians, were exceptions in this regard, for a number of reasons.

First, it was generally understood by knowledgeable people in both countries that Soviet theoretical physics and some areas of mathematics were at the very leading edge of world science, in second place to no one.[3] Thus, their American colleagues did not always need the direct assistance, and extra funding, of the formal exchange programs for their visits. They could and did organize these visits by themselves and fund them from their existing standard research grants or university funds.[4] Secondly, many of the top Soviet scientists, especially in theoretical physics as well as nuclear physics (both theoretical and experimental), were involved in classified military research, which brought along the usual restrictions common to many countries.

The year 1968 was very active in Thorne's field, gravitational physics. The Fifth International Conference on General Relativity and Gravitation took place in Tbilisi, Georgia, in September of that year. "This was a big event," Thorne recalled.

> I spent an afternoon with [John] Wheeler, Zel'dovich, and Sakharov in Zel'dovich's hotel room. It was just tremendously memorable. And I strengthened my ties to that Zel'dovich group there; to Novikov and Zel'dovich and the younger people in the group. And it was right after that that Zel'dovich invited me to come to Moscow to spend six weeks in Moscow. And so I went in '69 . . . on my own. There was an invitation to me from Moscow State University. I think it was initiated by Zel'dovich, although his name wasn't on it. . . . I didn't need anything else [e.g., support through the interacademy program] . . .
>
> Our field was one of the hottest fields in physics in that era. Basically, I became the conduit for information and ideas between the Russian and Western communities in this field. For me, that was the first of a set of visits that occurred roughly every other year from then until the fall of the Soviet Union in 1991. . . . [All these visits] were funded from my NSF [National Science Foundation] grants. In parallel, then, I had various Russians visit me through private invitations at Caltech. Novikov visited, Yevgeny Lifshitz visited once, Ginzburg came. All by private invitation. I never ever made use of the academy exchange programs.

We will hear more from Kip Thorne later about one of the most profound discoveries in the history of physics, in which as it turned out I played a very small supporting role.[5]

Tengiz Tsertsvadze

For many Soviet scientists, international conferences, like the one Kip Thorne attended in Tbilisi, were the first ticket to access to the world scientific community and to seeing the world. Tengiz Tsertsvadze, a biologist from Tbilisi, the very city that Thorne visited in 1968 for the conference on gravitational physics and relativity, tells of his first experience in the world of international science:

> The first contact, not cooperation but a contact, occurred in 1984. . . . It was a matter of chance and very unexpected. There was a World Congress of Immunologists and my professor, Academician Rem Petrov from Moscow, the Vice-President of the Academy of Sciences who was a very influential person, organized a group of fifteen scientists from all over the Soviet Union. Two scientists were from Georgia. He included us in this group. He organized a scientific study tour . . . to the scientific conference . . . in Montreal [followed by] visits to Washington, DC, and New York. The majority of scientists were from Moscow, but there were also scientists from Georgia, Armenia, Uzbekistan, and Latvia. . . . It was not cooperation yet. It was our first opportunity to visit the United States; the format was mostly tourism.[6]

The presence of so many Soviet scientists from outside of Russia was most unusual. This issue will come up soon in other testimonies. It was in part a harbinger of the promise of a new approach by Mikhail Gorbachev, who had just come to power but who had not yet launched his reformist agenda of glasnost (open discussion) and perestroika (rebuilding). But the delegation's cosmopolitan composition, as well as the idea of it in the first place, was Rem Petrov's personal initiative. Tsertsvadze described Petrov's motivation this way:

> He wanted to open a window between these two countries and make a first step so that Soviet scientists could attend the congress. Prior to that, it was impossible. When we returned, everybody envied us and could not believe that it happened. Academician Petrov used his influence in the government and, maybe, in other organizations to organize this trip. The main goal was to give us an opportunity. Now it sounds funny. Nowadays, anyone can purchase a ticket and go anywhere. My young employees travel to the US five or six times a year. They all have master's degrees from US universities. We also attended conferences in Europe and the US. It was impossible then. It was the first case when medical scientists traveled to the US. These were rare cases when scientists could travel abroad. It was a little easier to travel to Europe, but mostly to socialist countries, even more difficult was to travel to capitalist countries. It was impossible to go to the US before Gorbachev came to power. Under Gorbachev, Petrov felt the beginning of changes, a warmer climate, so to speak. He used this opportunity to organize the trip.[7]

Boris Shklovskii

Limitations on foreign travel, and even in participation in international scientific meetings held on the territory of the Soviet Union, existed across all scientific disciplines and across broad geographic areas. During the Soviet era, especially before 1985, Communist Party membership was a very strong determinant of whether one could travel to the West. Party members were considered more trustworthy. Geographic location also mattered a great deal; scientists who lived in union-republics of the USSR other than Russia, such as Tsertsvadze, generally traveled far less to the West than did Russians, and within Russia, Muscovites and Leningraders had favored treatment, in part because the Soviet scientific establishment was heavily concentrated in the two capitals.

Most affected by foreign travel restrictions, however, were Jews. Boris Shklovskii, the son of Iosif Shklovskii who was mentioned by Kip Thorne in his narrative, was a highly respected scientist in his own right, a recipient of a diploma certifying his membership in the Landau School of theoretical physics, which he proudly displays in his office today at the William I. Fine Theoretical Physics Institute at the University of Minnesota. In the 1970s, Shklovskii worked at the Ioffe Physico-Technical Institute of the USSR Academy of Sciences in Leningrad (now St. Petersburg). Unable because of his Jewish identity to travel to major international scientific meetings, Shklovskii attended smaller and more local scientific meetings—in his case, bilateral US-Soviet workshops—as his only available channel for direct interaction with foreign scientists: "Participation in international scientific meetings [in the Soviet Union] was a part of my experience in Russia, a very important part. I met quite a number of wonderful people and colleagues. . . . At those times when people were not visiting, the only interaction we had was writing letters. Letters were allowed, but they would, of course, have to be censored. And a letter in one direction could take three or four months. So it was very, very slow, but still it was important."[8]

In 1973, there was the first of a series of international meetings on solid-state physics held in Leningrad, organized by Western scientists to allow Soviet scientists who otherwise could not travel to the West—not only Jews—to attend. Shklovskii recalled:

> I was recognized by the Landau Institute of Theoretical Physics and that's why they invited me to participate in these theoretical symposia, several of them. I was part of one of them in Leningrad in 1973. Then the next time, probably

> 1976, it was in Moscow. I was not allowed by the local authorities to go to Moscow to participate in this meeting. Leningrad was a much stricter place [than] Moscow. The Leningrad [Communist Party] *obkom* [Regional Committee] and the KGB were stricter than in Moscow, and they didn't allow me to travel to this workshop. I don't know why—because they probably thought there will be too many Jews or something.... I do remember the one in 1979 at Lake Sevan.
>
> These were mind-opening experiences for me, particularly the one at Lake Sevan. The Afghanistan war had just started. Half of the American scientific community thought that they should boycott Russia, but another half fortunately felt that they shouldn't. The American delegation at that moment was chaired by Leo Kadanoff. And they did come to Armenia. I was allowed to go there. It was a fantastic event, particularly because we could hike in the mountains with our colleagues and then there was simply completely open interaction. And I benefited enormously from that interaction...
>
> And of course I'm enormously grateful to Leo Kadanoff, who died recently by the way.... I'm enormously grateful to him that he brought that group against the opinion of at least half of the scientific community. He was courageous enough to do that. And I appreciate this very much.[9]

Meetings such as the one at Lake Sevan were sometimes affairs at which small chinks were chiseled in the wall of security that the Soviets' "minders" from the KGB intended to impose: "Have you ever been to Lake Sevan?" Skhlovskii asked me when we spoke.

> It's a beautiful place.... So this is where we were located, in a Communist Party sanatorium, which was given to us to impress the Western scientists. The food was fantastic as always in Armenia, the wine and everything else. Except, it was prescribed for us where to sit during dinner, to sit at the same table the same chair, for the whole conference. In the registration file, everyone gets assigned a place at the table, the Americans too. The reason was that they [the KGB] had listening devices under the table, and for them, it was important that we sit in [assigned seats] because voice recognition was not great.
>
> We ... didn't [really] mind; we were happy just to be there. But when the Americans arrived and got their registration files and saw that they had assigned seats, they said they would boycott the dinner. So they didn't even show up for dinner.... So we [the Soviet scientists] ate our dinner, and we understood what [was] happening. Then what happened was that this was reported back to Moscow, to the KGB headquarters or whatever. It was a scandal because [the KGB] knew that the Americans just barely showed up at all because of Afghanistan, so if they did not lift the requirement, the Americans would probably leave. And this would be a bigger scandal for everybody who was involved in the Russian side, and eventually we got permission from Moscow to allow everyone to sit anywhere they wanted. So the next morning, we got pieces of paper saying that the seating requirement will be ignored. It's a small American story, but it impressed us enormously.[10]

It was not until 1989 that Shklovskii was allowed to travel to the United States, and that was with a one-way ticket from the University of Minnesota. The University of Minnesota invited him to head a new institute there and offered him a generous discretionary fund with which he could bring Russian scientists to the Twin Cities for short- and long-term visits. Some of the most well-known names of Soviet physics were his guests. So overwhelming was the number of distinguished visitors that for a while, the University of Minnesota was known in the physics community as "Moscow on the Mississippi."

Roald Sagdeev

International conferences, as in Shklovskii's story, were often occasions for strange and wonderful things to happen. The young Roald Sagdeev had just started working in the "Bureau of Electronic Equipment" at the "Laboratory of Measuring Instruments." This was the code name for the enormous, classified institute headed by the famous Igor Kurchatov and that later became the Kurchatov Institute. Sagdeev, among others, was picked by Kurchatov in 1958 to attend a major international meeting on peaceful uses of atomic power in Geneva, his first international experience. The meeting was held in association with the "Atoms for Peace" exhibit, which, along with the initiation of scientific and other exchange programs, was a key feature of President Eisenhower's overture to the Soviet Union. The subject of the scientific conference was controlled nuclear fusion.

Sagdeev recalled arriving in Geneva and walking the streets with his friends. "The reality we encountered," he wrote in his autobiography, "was not simply charming, it disarmed us. As we encountered people walking along the streets of Geneva, we experienced a strong shock. They were without the imprint of fear on their faces. They looked prosperous; indeed, some looked wealthy. . . . We started asking each other 'Where is the oppressed working class? Where are the proletarians?'"[11] This was a common reaction for many Soviet scientists and others arriving for the first time in the West, especially before the late 1980s. To be sure, many Americans experienced a similar kind of shock upon first setting foot in the Soviet Union; they might well have asked, "Where is the liberated proletariat? Where is the New Socialist Man?" It was a strange encounter for the people of both worlds.

Sagdeev then related how a member of the Soviet delegation, Professor Konstantinov, attached himself to his group and made himself particularly evident whenever there were conversations with the Americans. His English was far better than any of the Soviet scientists'. In one encounter, when Sagdeev and others went to a movie with a young American scientist, the mathematical physicist Martin Kruskal, Konstantinov made a drunken spectacle of himself with political speeches and later scolded the young Soviets for not beating up their new American friend. Sagdeev and the others decided to tell the story to . . . one of the senior scientists on the Soviet team, who told it to the Kurchatov Institute's Communist Party chief, who in turn relayed it to higher authorities. Within a few hours, Sagdeev and his friends found themselves in a grand suite in Geneva's Metropol Hotel, where they were introduced to Ivan Serbin, whom Sagdeev describes as "the gray cardinal of the Soviet group here." Serbin thought for a while and then he spoke:

> "Don't worry. Continue your duties at the conference. Nobody will undermine your efforts in the international arena anymore." He rose from his chair, indicating that the conference was over. . . . We did not have to wait long for the results of that meeting. By lunchtime Konstantinov, seemingly embarrassed, approached us and said a few words, trying to apologize for his misbehavior or the misunderstanding. . . . With these words, he handed us a small sum of money, a few Swiss francs, as reimbursement for last night's tickets to the movie. That was all. But it was a symbol of our victory.[12]

"The rest of the conference," Sagdeev wrote, "was even more exciting." While it "brought very few ideas from each side that had not been independently discovered by the opposite team," it became clear that the Americans had thought of some approaches that the Soviets had not. There was some embarrassment when Lev Artsimovich, the team's most senior member, tried to argue that the Soviets had done comparable work, but Mikhail Leontovich spoke up and clearly stated that this was not the case. "His intervention," according to Sagdeev, "was a remarkable lesson of scientific integrity—and, I would say, even human decency—for us young scientists. After all, at the end of the conference both teams agreed in a somewhat teasing way on the unofficial score of three to one in favor of the Americans in experimental plasma and fusion physics and one to one in theory. That made all of us who were Leontovich's pupils extremely proud."[13]

Marjorie Senechal

In mathematics as well as theoretical physics, there were strong intrinsic scientific reasons for US mathematicians to get to know the work of Soviet experts. Marjorie Senechal, of Smith College, was attracted by the unique approach of Soviet scientists to the topology of crystals:

> I got involved through a scientific problem that I was interested in; it had to do with crystals and with their shapes, and is a problem that still interests me.... I was fascinated by the crystal shapes, and the question is, why do they have the shapes that they do? The usual answer is that their atoms line up in rows and layers, like building blocks, and approximate these beautiful shapes. But some crystal shapes are impossible to explain that way. For example, you sometimes find crystals that look like two cubes growing right through each other.... So there has to be some way of understanding how something like that can form. Crystals like these are called twins.... So I began studying and reading the crystal twinning literature. Even in English it made no sense to me whatsoever; not that I couldn't understand it, but I didn't believe it.... I worked on this very hard and I wrote a paper about it ... but I still felt that there was something missing from my understanding.[14]

Senechal found what she was missing in the work of a Soviet geometer whose work was published only in Russian in Soviet scientific journals:

> A student who was working with me at the time found in a paper that she thought might interest me in a journal of the [USSR] Academy of Sciences. I read this and suddenly I saw how the twinning might work, and it got me very, very excited.... I really [wanted to know], who is this person in Russia? It turned out to be Nikolai Naumovich Sheftal at the Institute of Crystallography. So I wrote to him and I asked him some of my questions about this and said how much I liked his paper and asked if he thought it would work in the case I was interested in, which was more complicated. He wrote back wonderful encouragement to me in reasonable English, and we carried on this correspondence.[15]

But correspondence was no substitute for establishing a close working relationship. Senechal discovered the opportunity to make that happen through the interacademy exchange program:

> Then I found out about the interacademy exchange program. I thought, "I would love to work with him." That's somebody I really valued, and so I applied for that. And I also applied to work with somebody else who was doing more mathematical work but also crystallography. The second one turned out to have been a mistake, that was the professor I kicked out [of my office] near the end of my stay...[16]
>
> I got the interacademy exchange program award and to my great delight, they gave me eleven months, which I thought was amazing, and my whole family could go, including two school-aged daughters. And so we went; they went to school there, [and] I worked in the office and lab with Sheftal for nearly a year.... And so that's how it got started.[17]

And it went on from there. Senechal not only continued her collaboration but also became deeply involved in the governance of nonprofit exchange and cooperative programs with the Soviet Union and later Russia, through the NAS and through the Civilian Research and Development Foundation (CRDF), and she cochaired a major program to introduce deep structural and cultural change in Russian universities, CRDF's Basic Research in Higher Education in Russia program.[18]

Lawrence Crum

Larry Crum, research professor of electrical engineering and bioengineering at the University of Washington, was already an active participant in the international scientific community through the vehicle of international scientific meetings when he first became aware of very advanced Soviet work in acoustics.

In the mid-1980s, there was a meeting of the International Conference on Nonlinear Acoustics in Oslo. Crum recalled:

> The Russians sent a ship over to Oslo with fifty or seventy-five Russian scientists. That was cheap for them because they could live on the ship and attend the meeting. I met a young lady there, and she gave a remarkable talk. She was friendly, and she was outgoing, and she said, "I would love to come to the United States to collaborate with you." Eventually she came, and her colleagues came, and we established a relationship that lasts to this day. I think that the first occasion of her visit was through the CRDF program, through which we sent money to them and then they came here for a couple of weeks.[19]

Crum explained why Soviet acoustical science was of such intrinsic scientific interest. In a field like nonlinear dynamics, because they could not crunch numbers as efficiently as others, they started from deep theoretical analysis:

> It turns out that in Russia, because they have to do a lot of the work theoretically; they invented nonlinear acoustics. They have their names on the equations, and so when medical ultrasound started getting into high-intensity ultrasound, we had to learn how to do nonlinear acoustics. These were the people who knew how to do nonlinear acoustics. In this scientific meeting in Oslo, these people came and those Americans who were doing nonlinear acoustics were saying, "Oh my God, these people are ten years ahead of us, and we need their information if we're going to develop very practical sorts of US technology!" The Russians wanted to develop it in an intellectual sense.... We wanted to apply that technology to humans, and we did.[20]

Advanced acoustical science, of course, has more than biomedical uses. Indeed, in my years of coordinating exchange visits and joint research projects with the Soviets and later the Russians, there was no field that was more sensitive and of greater concern to the US military and intelligence community than acoustics. In our discussion, he offered an example of just how advanced their work was, both in theory and in practice: "Here's a bit of a secret; it's not classified, but it's not commonly known. I did research on wake-calming torpedoes. It turns out one of the astronauts looked down and saw that whenever a ship was out he could see a wake that would go on for miles. And so the Russians decided that they could fire a wake-calming torpedo, and it would go into the wake, and it would just follow the wake and would run up until it hit the propellers of the aircraft carrier and destroy the aircraft carrier."[21]

There's some pretty fancy nonlinear hydrodynamics and acoustics that go into an accomplishment like that. This was a military technology that the Soviets had but that we did not, at least at the time. It is easy to understand, then, why this entire area of research should be so closely guarded by both countries. Yet its roots were not in technology but in the most basic scientific research imaginable: the modeling of chaotic phenomena. And it was not "copied" or "stolen" or "reverse-engineered" technology. It was military technology rooted in the advanced work of Soviet mathematicians in theoretical and computational mathematics. This Soviet basic research made it possible for their scientists and engineers to have extraordinary skills in mathematical modeling. In part, it arose from scientific traditions specific to Russia, and in part from their lack of access to the kinds of immense digital computing that existed in the West.

This is a pattern that we will encounter frequently throughout this book. In practice, it set the framework for some intense discussions between the scientific and government communities in the United States—and no doubt in the Soviet Union as well—about the limits of unrestricted international scientific cooperation.

Yaroslav Yatskiv

In some areas of science, like astronomy and space research, international cooperation is organic to science itself, especially where very large instruments, coordinated observations from widely separated locations, and complex missions into outer space are concerned. For Yaroslav Yatskiv, the director of the Main Astrophysical Observatory of the National Academy of Sciences of Ukraine, the international scientific societies were the ticket to becoming involved in international science cooperation: "I was very lucky in life regarding scientific collaboration in that I had a teacher, Professor E. P. Fedorov, a world-renowned scientist in the field of astronomy and Earth rotation, namely determination of the Earth's rotational parameters, and the so-called nutations of the Earth's axis."[22]

The Soviet Union, of course, was a major actor in outer space. The Vega program of 1984–85 to investigate Venus and then Halley's comet was a huge multinational effort in which the Soviet Union had a leading role.[23] Yatskiv became a part of that effort.

> I was lucky when Roald Sagdeev [at that time, director of the Space Research Institute of the USSR Academy of Sciences] needed a deputy and assistant for the determination of the orbit of the spaceship *Vega* and Halley's comet. The space project Vega was the most successful, most recent project on Halley's comet. The Europeans had the spaceship *Giotto*, the Americans had no spaceship, but participated in the so-called Inter-Agency Consultative Group for the exploration of Halley's comet. Sagdeev invited me to be his deputy for navigation and determination of the position of Halley's comet, so that the spacecraft *Vega* could approach Halley's comet at a distance of ten thousand kilometers. Yes, yes, that was an achievement then. This was the beginning of the eighties. Now *Philae* has even landed on a comet from the spacecraft *Rosetta*, but then it was very difficult.[24]

Peter Raven

In other areas of science for which geographic location is important—earth, ocean, atmospheric, botanical, and zoological sciences, and so forth—it was not necessarily the excellence of the science itself in the Soviet Union but rather unique locations or collections that were the attractors for scientists from the United States and elsewhere. For Peter Raven, it was both. He described his first encounter with Russia, in Russian California:[25]

> My first sort of professional recognition that there was a Russia was . . . Russian California. The California poppy—the state flower—was first collected by Russians in or near where the Presidio was built in San Francisco during a stop on an expedition down the coast. Its scientific name is *Eschscholzia*. . . . A beetle that I used to collect in the Santa Cruz Mountains was *Nebria eschscholzii*. Russian scientists collected the first specimens of that particular beetle on a biological expedition [from] Fort Ross to the interior of California. So I had come to know that they had been there . . .
> I have always been interested in countries around the world. In 1956, when I was an undergraduate student at Berkeley, I took a course on the geography of the Soviet Union, taught by Professor Nicholas Mirov, a Russian émigré forester. I found the huge size and diversity of the country interesting and began thinking about it more than I had earlier, learning about it.[26]

As for so many other scientists, Raven's first contact with Soviet Russia was through a scientific conference, which was held in Leningrad.

> The first chance for me to go there was the International Botanical Congress in 1975.... That year, the world's botanists were invited to come to Leningrad—a very novel opportunity at the time. Contacts earlier had been few, and sometimes dangerous. On that occasion, we stayed in the country for nearly four weeks.
> Several years earlier, 1960–61, I had spent a postdoctoral year in London, concentrating on the study of a group of plants known as willow weeds, *Epilobium*. At the time, I learned of a leading Russian specialist on the group, Alexei Skvortsov, who was at the Main Botanical Garden in Moscow. When I finally met him at the congress, he told me that if I'd even written him a letter in 1960 he quite possibly would have been jailed and conceivably even killed. That made me awfully glad that I had not done so, even though I couldn't have imagined such consequences. Alexei may have been exaggerating, but he was certainly expressing real concern. Even at the congress, contacts were rather limited and public.[27]

Raven's interaction with Skvortsov in 1975 was typical of the times: "He gave me a manuscript to publish in the US. And the way he gave it to me was that we took a ride on a bus. We were doing that all the time. I took several rides on buses with people to talk. Nothing secret, you know, everything was always recorded, but the thing is I was always hoping I'd say something interesting enough for anybody to care about recording it."[28]

During the 1975 International Botanical Congress, Raven had his first opportunity to visit the famous Komarov Botanical Institute of the Soviet Academy of Sciences in Leningrad. This was the beginning of his long infatuation—perhaps better phrased as obsession—with the Komarov Institute.

> We met in Leningrad for about ten days, and I became familiar with conditions at the world-class Komarov Botanical Institute. A number of the botanists visiting that year noticed that chunks of the plaster were falling off the walls, and leaks were wreaking havoc on the exterior of the building. We learned that there was not enough heat for people to work there in the winter: the boilers were broken. The building at that time was only about sixty years old, but it had been poorly constructed. Several of the delegates to the Congress noted that the building was in poor condition and wrote articles about it when they returned to their homes.
> It seemed a tragedy to the whole field of botany, because within it were housed the major collections of plants from the whole huge area of the Soviet Union, a seventh of the world's land surface, as well as many important collections from China, Latin America, and elsewhere where Russian botanists had ventured over the years. Many leaders in the field of plant classification, systematics, have been Russian, and that gives extra importance to the specimens

that they studied. The Komarov holds one of the largest and most important study collections of plants in the world, an asset that is both unique and irreplaceable. For this reason, the poor condition of the building in which those collections were housed was a matter of major concern to those who participated in the 1975 congress.[29]

At the time, there was little the Western botanists could do about the institute's condition aside from developing and maintaining scientific contacts with them. But seventeen years later, when George Soros appeared on the scene with his International Science Foundation, Raven recognized a unique opportunity to act and enlisted Soros and others in a million-dollar project to repair the institute's roof.[30]

The botanical congress was extremely productive in other ways too. There were field trips to relatively remote areas of the Soviet Union, allowing Western botanists access for almost the first time. Raven chose a trip to Georgia.

> In connection with the congress, the Soviets had really opened up their country to their visitors by organizing a wide-ranging series of excursions. After a good visit to Moscow and its surroundings, we joined a weeklong trip to the Republic of Georgia, which would have seemed unimaginable just a few years previously...
>
> After spending several days in and around Tbilisi, we headed to Lagodekhi National Park, which contains one of the most beautiful and biologically rich areas in the Caucasus, a temperate forest that is filled with unusual plants. That early visit helped to point the way to a great deal of involvement by the Missouri Botanical Garden in the Caucasian region over the past twenty years, with both botanical activities, ones dedicated to conservation, and horticultural exploration. Some of the plants that Alexei Skvortsov and I had studied were there, and the trip was a great thrill to me.[31]

Karl Western

Karl Western, a senior adviser at the National Institute of Allergy and Infectious Diseases and a longtime manager of international cooperative programs at the NIH, had his first exposure to Soviet science in an unusual location: Nigeria, during the horrific Nigerian civil war of the mid-1960s. At the time, he was a young staff scientist at the Centers for Disease Control in Atlanta, Georgia.

> At the time of the civil war, most of the trained Nigerians, compared with today, were relatively few [and] were from Biafra, which had seceded, so they took with them most of the officers who were engineers and physicians and public health people.... The political divide in the Nigerian civil war was

that the French, the Spanish, the Portuguese, and the Chinese were strongly supportive of Biafra because it was secessionist, and the friends of the Federal Government of Nigeria were the UK, US, and Russia, or the Soviet Union. Russia had a very big stake in the smallpox eradication program [because they were working on a smallpox vaccine]. . . . Many of the aid staff were Russian, so we had Russians that were working on the West African smallpox program. That was my first exposure to them.

I didn't have any very close relations with them, but one little story was that the English and the US didn't want it to be known as the Russians were active, so they gave most of us codenames. So Dr. Isimov . . . would not be Isimov, he would be Dr. Jones. [For my codename,] they thought "Western" was good enough. There were two Americans to whom they did not give codenames: Larry Brilliant and me. They thought "Dr. Brilliant" and "Dr. Western" sounded codename enough. And the real purpose was to disguise the fact that the Russians were actively partnering with us.[32]

Contacts between Soviet and American scientists at this time, especially working relationships, outside of either the official exchange programs or scientific meetings were extremely rare indeed. To be sure, this is a channel of cooperation that is not uncommon with other countries, but the politics of the day made it very unlikely for this to happen between American and Soviet scientists.

Julie Brigham-Grette

For Julie Brigham-Grette, professor of geosciences at the University of Massachusetts–Amherst, it was the Bering Land Bridge—and the unique physical environment of the Soviet Union—that attracted interest:

When we first saw the Soviet Union starting to crack, my mentor, David Hopkins, who is like Mr. Bering Land Bridge . . . had very good connections in Russia—though he had never been there. . . —to Chukotka. He had written books about the Great Land Bridge and had always wanted to get over there. So he and I wrote a grant proposal to NSF to do shared science, where we would go to Russia and look at their sites and then we would bring Russian scientists over here. . . . The funding came in about 1990 and then our first trip was in '91 and '92. . . . We were working with a group out of Magadan and this was where . . . Dave Hopkins had good connections.[33]

At a deeper level, it is important to understand why visiting the Soviet Union was of relatively intense interest to American scientists in what one might call the "data-dependent" natural sciences: botany, zoology, geology and geophysics, archaeology, atmospheric sciences, oceanic science and limnology, paleoclimate studies, and more. Here, the magnet was not necessarily the quality of the science done in the Soviet Union or certainly the volume of publications in the international literature, of which there were very few because Soviet scientists in these areas published almost exclusively in Soviet journals in the Russian language.

The powerful attractor was the location. The Soviet Union accounted for one-seventh of Earth's landmass. In its territory were many unique species, biomes, geological formations and minerals, meteorological and atmospheric conditions, and records of ancient cataclysmic events. Access to these sites was complicated, both logistically and especially politically. For these scientists, the formal exchange and cooperative programs, with their "roofs" (in Russian *kryshi*) of political and diplomatic shelter and their networks of organizational support from high-level organizations such as the Academy of Sciences, were often essential. The geoscientist who tried to do it on her own, like Brigham-Grette, sometimes encountered seemingly overwhelming difficulties. Her story of "the expedition from hell" in chapter 11 should give pause to any "data-dependent" scientist who fancies that she or he can simply organize a field expedition in Russia or Central Asia like one would in France or South Africa.

Roald Hoffmann

Not everyone who entered the realm of US-FSU science cooperation, however, did so straightforwardly. Two curious stories of this type are told by Nobel Laureate Roald Hoffmann of Cornell University's Department of Chemistry and Loren Graham of Harvard University and the Massachusetts Institute of Technology. First, Hoffmann, whose introduction to the Soviet Union arose out of his desire to *avoid* making a commitment to chemistry at the time relates his story:

> I was born in 1937, in what was then Poland, and then became the Soviet Union and now it's Ukraine. My mother and I and my stepfather (my father was killed by the Nazis in 1943) came to the United States in 1949. I was eleven and a half at the time. We settled in New York City; I went through the usual public schools, and then on to Columbia and to graduate school at Harvard. My beginning of involvement with the Soviet Union was in the academic year 1959 to 1960, which was my second year in graduate school at Harvard. So at this point, I'm twenty-two to twenty-three years old. I had leftist inclinations typical of Jewish immigrants at that time (or perhaps the prewar period), but I was certainly not a Communist.
>
> I found out about the IREX exchange program in 1959. There was a scientific factor in thinking about the Soviet Union, and that was a visit to Harvard by a leading physical chemist named Michael Kasha, who was of Ukrainian

origin and who taught at Florida State University in Tallahassee. Kasha taught a course about energy transfer in molecules, and in that course, he talked approvingly of the work of two Soviet scientists, Terenin and Davydov. Alexander Sergeevich Davydov, the person I eventually wound up working with at Moscow University, had done something very important. He had studied in Leningrad with Ioffe and had developed a workable theory of what were called excitons, describing how energy migrated in crystals.[34]

Yet even though Hoffmann had become interested in the work of Soviet scientists, he was unsure about a career for himself in science at all at the time:

> I'm a rather unusual chemist in that I did not make a commitment to chemistry until three quarters of the way through my PhD in chemistry! In my first two years at Harvard, I sat in on courses in other fields. In astronomy, modern science policy . . . I think I was uncertain about chemistry. That uncertainty originated in college, where I finally worked up enough courage to tell my parents I wasn't going to be a real doctor—there was a lot of pressure to go into medicine. At the same time, the world opened up to me at Columbia, in arts and literature, and the science teaching was uninspiring until my last year. I didn't have enough courage to tell my parents that I wanted to study art history. That would have killed them. Anyway, I went to graduate school in chemistry as a kind of compromise. I did well in courses, but my heart was not in it—not yet.
> So I think the interest I took in going to the Soviet Union in my second year of my PhD program in chemistry at Harvard, applying for the IREX program, was, in part, a procrastination, a delay in making commitment to chemistry. As it happened in the Soviet Union, I worked with Davydov, a physicist in the physics faculty. But I always was in between chemistry and physics, and still to this day, even as my heart is in chemistry, I do things that are close to physics. Also, in the summer of 1959, I went to Europe for the first visit there since our coming to the United States. I went to a summer school in Sweden. I met my future wife, and in the middle of that year, we decided to get married. That was just around January of 1960. And at the same time, there was a deadline for applying for this IREX program.
> As part of the application process, I was interviewed by a committee headed by Marshall Shulman,[35] I think, of Columbia University. They were suspicious of my motives. Especially since I told them that I'm getting married to a Swede; that seemed strange. Anyway, they did admit me to the program; then my wife, Eva, came over in April, we got married at the end of April. We went to study Russian at Indiana University—in the program that many people took part in in those days, an intensive program in the summer of 1960. And in September, we went to the Soviet Union.[36]

Together with Graham, Hoffmann thus became one of the first scientists to spend a year in the Soviet Union. This is noteworthy because the visit took place under IREX and not the brand-new interacademy program, which sponsored only short-term visits of up to one month and, for that matter, only for senior scientists, not graduate students.

> The IREX program began in 1959. I was in the second year of the program. In that year, 1960–61, I was one of two scientists out of the forty US exchangees. And the Russians sent all scientists and engineers. The other American scientist was Ole Mathisen, originally from Norway, who was at University of Washington, He had the most wonderful time of us all because he was an expert [in] salmon fishing. The Russians were very interested in salmon fisheries and the science of salmon. So he got to travel much more than we did. There were a number of people there. Loren Graham was one of them. Many people we sent over were Slavicists and historians. They were continually goading the Soviet bureaucracy to allow them access to this or that archive. The Soviets resisted.

Indeed, it was the IREX program's mixed nature, including both natural and social scientists, that made it more "political" in a sense than the NAS program. The IREX program was constantly beset in these early years, especially, by struggles to maintain "reciprocity," which translated into balance between the number and duration of visits by natural versus social scientists. The Soviets wanted to send more natural scientists; the Americans wanted to send more social scientists and historians. Of course this was a function of the interests of their scholarly communities and their countries' political priorities. Hoffmann continued:

> We stayed that year in Moscow in that wedding cake of a building of the university. I worked with Davydov. It was a wonderful year in many different ways. Certainly, my depth of knowledge of Soviet culture and the language grew in that period. It was also our honeymoon! I was at this point in 1960, just twenty-three years old, a good age. It was my third year in graduate school. Harvard thought I was crazy to go. They tried to dissuade me. My parents didn't like it at all. They thought I would be drafted, because I was born on the territory of the Soviet Union.[37]

Loren Graham

Loren Graham, professor of the history of science at MIT, who was in the same group of US exchangees as Hoffmann, had already worked as a chemical engineer in the 1950s but, somewhat like Hoffmann, had doubts about a career in that field:

> I have a degree in engineering and I worked as a research engineer for a short while for the Dow Chemical Company. I found out there that I love science and technology, but I didn't want to do it. I wanted to write about it. I wasn't meant to spend my life in a laboratory. I was meant to spend my life at a writing desk. And the way to do that, keeping my interest in science and technology

and yet being able to write, rather than to work in a laboratory, was to go into an academic field that might permit that—and that was the history of science.

So I went into the history of science and I got a PhD combining Russian history with history of science. Why Russian history? Well, I entered graduate school in 1958. Remember that *Sputnik* went up in 1957. So there was a sudden great interest in the West in Russian science and technology, but almost nobody knew anything about it in its institutional, political, and social setting, not to speak of its history. We were amazingly ignorant in 1958 about those things. So, I said, "Hey, that's my thing!" and I started studying the Russian language. In order to get a PhD at Columbia, you had to have two foreign languages, so I presented French and Russian. I read French well, but my tongue—my midwestern tongue—was not constructed for French. But it turned out that my midwestern tongue is not constructed that badly for Russian. I fell in love with the Russian language and I still love the Russian language. So that's how I got into it.[38]

Loren Graham went on to a distinguished career as the premier living Western historian of Russian and Soviet science. He fostered, and indeed created, an entire discipline of study in the United States of the study of the role of scientists and engineers in Soviet and post-Soviet society.

Siegfried Hecker

For many scientists, if not all, their stories about how and why they got involved were mixed, as one might think. The mixes, however, each had their own individual signature. Sig Hecker, the director of Los Alamos National Laboratory from 1986 to 1997 recalled:

> It was because of the science. . . . It was because these guys invented the radio frequency quadrupole. . . . The Americans took it. Ed Knapp, who later became [NSF director] . . . is the guy who took the radio frequency quadrupole and put it into the lab facility and so these things are built on each other, so then we knew the Soviets. . .
>
> So when I went there as director of Los Alamos National Laboratory— instant respect. Los Alamos was the nuclear mecca of the world, and it was not only because we built the bomb; it was they know all the scientists who came from Los Alamos. So . . . that actually gets us to the time when we first met in '88, when it was sort of love at first sight. We saw the scientists. We . . . didn't see the Russian nukes staring at us; we saw the scientists.

In their early scientific collaborations during the Soviet period and even later, Hecker said,

> most important turned out to be an idea that goes in Russia all the way back to Sakharov and what he called magnetic accumulation, essentially the compression of magnetic fields explosively. . . . The Russians were good at that. Sakharov had the initial idea. The people at VNIIEF—they were in Los Alamos—they

carried it on and then we had a parallel program at Los Alamos under a fellow by the name of Max Fowler. These guys got together during what were called Mega-Gauss conferences; the first one goes all the way back to 1965 in Frascati, Italy, and there [have] been a number since. So we had this sort of scientific collaboration, but they were specifically targeted in a couple of areas and especially in things like the magnetic compression accumulation business.[39]

Terry Lowe

Terry Lowe, research professor of metallurgical and materials engineering at the Colorado School of Mines, had worked at Los Alamos closely with Hecker in the lab's Metallurgy Division prior to Hecker's promotion as director in 1986 and later became deeply involved in nonproliferation projects with Russia both as a Los Alamos scientist and later as an entrepreneur. Lowe had also taken an early interest in unique Soviet research, in this case on nanomaterials, which at the time was just becoming a hot topic in the West:

> In the United States we have an organization called Materials Research Society, MRS. And there's a European version of that society, the EMRS. They were having their annual meeting in Saint Petersburg, Russia in 1992, just after the fall of the Soviet Union. I decided to go to that meeting and meet Russian scientists, and I had the opportunity to interview and meet with literally hundreds; there were 450 scientists there for that meeting...
>
> Earlier, I had been at a National Science Foundation meeting in San Diego, at the University of California in San Diego, where I met a man named Mikhail Zelin. Michael Zelin had come to the United States as a scientist and was at this meeting. I described my interest in validating the supercomputing models that we use to study the behavior of materials, particularly trying to look at nanoscale phenomena. These are 10^{-9}-meter phenomena; and at the time, it was virtually impossible to do experimental validation of the simulations since they could only run for very small amounts of time—think picoseconds to nanoseconds—and only simulate behavior of very small volumes of material.
>
> So here was this Russian scientist who said that he could make nanostructured materials, that is, nanoscaled materials. He mentioned an institute in Russia, the Ufa State Aviation Technical University, and his former advisor, whose name was Ruslan Valiev, and who was the person who was developing this technology. And I said, "Well, that's very interesting. This technology might enable us to do experimental validation of these large-scale simulations that we were conducting at Los Alamos."[40]

Lowe's entry into the area of nonproliferation work with Russia came later:

> From 1990 until 1996, I was the group leader of Materials, Research and Processing Science at Los Alamos National Laboratory. That organization was one of several in the Materials Science Division, of which Sig Hecker was the

former leader. My office was in the Chemistry and Metallurgical Research Building (the CMR Building), where I used to have a locker next to Sig's. That is where this all started, in essence, as I listened to him, who at the time was the lab director of all of Los Alamos National Laboratory. Talking about the consequences of the fall of the Soviet Union, he said, "Hey, we need to get over there and do something." I don't want to say this was prior to any thought in Washington, DC, along the same lines, but it was in part independent of those thoughts, and so the question was what to do.... I actually took the initiative to go to the former Soviet Union to meet with Russian scientists.... It was really inspired by Sig. It's because of him that I made that first outreach.[41]

As Hecker and Lowe both make clear, it was the scientific interest and connections as fellow scientists, not policy considerations, that preceded their very applied nonproliferation work after 1991. In the latter turbulent years, the purely scientific collaborations continued as well. Hecker devotes an entire chapter of his recent edited two-volume book, *Doomed to Cooperate*, to the basic research carried out by the US lab scientists (not only from Los Alamos) and their Russian colleagues in the closed nuclear cities.[42] As we will see later on, these research collaborations were intimately linked to, and essential to, the nonproliferation work that proceeded in parallel and gained most attention from the global community.

The Diplomats

A second common path toward the sphere of science cooperation was through diplomatic service. I include here both career diplomats whom I had the privilege of interviewing, as well as individuals who spent significant parts of their careers in diplomatic service but went on to make their names in other capacities and organizations. The latter group may not self-identify now as diplomats, but for me, as well as I hope the reader, it is helpful to designate them as such for the purpose of understanding how they initially engaged in the business of US-Soviet science cooperation, and why.

Thomas Pickering and Eileen Malloy

Little personal explanation is needed for how career diplomats like two former ambassadors I had the privilege of interviewing for this book, Thomas Pickering and Eileen Malloy, intersected with the sphere of scientific cooperation. Pickering's role in this area dates back at least to the period 1978–81, when he served as assistant secretary of state for oceans and international environmental and scientific affairs; earlier, he had been US ambassador to

Jordan and before that, a special assistant to Secretaries of State William Rogers and Henry Kissinger. Later, as ambassador to Israel, India, and Russia (1993–96), he was deeply aware of the significance of bilateral science cooperation as an element of US foreign policy and, indeed, was always highly supportive of and well informed about such efforts. After his retirement from diplomatic service, he became vice president for international affairs at Boeing and a passionate spokesman for international scientific cooperation. Malloy, US ambassador to Kyrgyzstan from 1994 to 1997, had earlier served in the Environment, Science and Oceans Office at the US embassy in Moscow.

Glenn Schweitzer

Glenn Schweitzer, the long-serving director of scientific exchanges programs with the FSU and eastern Europe at the National Academy of Sciences, also got his start in the area through diplomacy. In 1962, he recalled in a personal memorandum:

> I was on assignment to the Science Division of the newly established Arms Control and Disarmament Agency. The Department of State's Office of Personnel telephoned me to inform me that as a regular Foreign Service Officer, I was scheduled for my second overseas assignment and I would soon be contacted by an office interested in my background. The next day the Soviet desk of the Department invited me for an interview. The desk officer informed me that I was being considered for assignment to the Embassy in Moscow as the first Science Officer. While there were Science Officers at about twenty other embassies, all of them had been recruited from universities or industry; but he underscored that Moscow was a special case. They needed someone in Moscow who not only had a good background in science and technology but whom they could trust to follow strictly instructions from the Department and the Embassy. Their solution was to assign a career Foreign Service Officer to the position, and they had discovered that I not only had the required background, but I had a degree of fluency in Russian. I immediately accepted the offer of an assignment and within six months I was on my way to Moscow.[43]

Through this assignment, Schweitzer became the first science officer ever to serve in US embassy Moscow. It also thus became a pattern to assign career foreign service officers to this important post, as opposed to scientists, in part in recognition of the unusual political importance of scientific ties with the USSR. Schweitzer did not long remain in the Foreign Service, going on to a broad variety of posts in government and nonprofit organizations. His story about how he came to work at the NAS in 1985 appears in chapter 9.

Norman Neureiter

For Norman Neureiter, too, the Foreign Service was an important stop on the critical path to his lifelong engagement in international scientific cooperation. In his case, he had already substantial experience as a research scientist at Humble Oil and science administrator at the NSF before joining the Foreign Service in 1965. He was then immediately assigned to the US embassy in Bonn as deputy scientific attaché, and two years later named as the first scientific attaché at US embassy Warsaw, with regional geographic responsibility for Poland, Czechoslovakia, and Hungary. In subsequent roles at the White House Science Office; as vice president for international affairs at Texas Instruments; science advisor to Secretary of State Colin Powell; and in other roles, Neureiter had a major impact on scientific cooperation worldwide and is today one of its most passionate advocates.

We will look into Neureiter's detailed narrative a little later in this chapter, when we consider another powerful magnet drawing people, including myself, into the field of science cooperation with the Soviet Union: fascination with Russian language and culture. For now, however, we move from engagement through diplomacy to engagement through another policy avenue more specific to the post-1991 period: preventing the proliferation of weapons and technologies of mass destruction.

Andrew Weber

Andrew Weber and Siegfried Hecker are probably the two most celebrated figures in the historic effort to combat the nonproliferation of weapons of mass destruction from the FSU at its source. Weber's debut in this field was as the head of the political-military section of the US embassy in Alma-Aty, Kazakhstan. One day he told me:

> It started very specifically with an automobile mechanic, who asked me if I wanted to buy some uranium, and this is in David Hoffman's book, *The Dead Hand*,[44] so there's not much to add here. But to Washington, it looked like something very far-fetched. There was a lot of skepticism. There were a lot of scams occurring at that time in history—red mercury was a big one—but I felt it was worth pursuing, even if there was nothing there. I developed relationships with the factory director where this material was allegedly stored and then eventually found out from him how much material and how highly rich the material was, and we arranged with President Nazarbayev for a quiet visit to the facility to confirm the existence of this material. That was in March of 1994, a very, very snowy day, and we went up for a visit with a technical

expert from Oak Ridge National Laboratory and indeed we confirmed that there were approximately six hundred kilograms of 90 percent enriched HEU [highly enriched uranium].[45]

This material came from the Ulba Metallurgical Plant in Ust'-Kamenogorsk, Kazakhstan, where there were rich deposits of uranium and other rare earth metals. The Ulba factory "produced LEU pellets, but they had, in the Soviet era, produced HEU beryllium alloy fuel rods for an experimental submarine program called modular nuclear reactors. The materials were just sitting there, and Moscow had more or less lost knowledge of them, or the records were on paper somewhere, but very few people knew that this much material was there. The whole issue of vulnerability to theft and the black market was new, and these had been the closed cities where nobody had access."[46]

Weber had the Ulba uranium tested and got confirmation that it was indeed the highest grade of highly enriched uranium used to make nuclear bombs. In what came to be known as Project Sapphire, all of the HEU was shipped back to the United States and reprocessed at the Y-12 facility at Oak Ridge, Tennessee, into low-enriched uranium (LEU) not suitable for weapons use.[47]

But that was just the beginning of Weber's story. Kazakhstan was the site of production not only of HEU for Soviet nuclear weapons but also of virulent offensive biological warfare agents. Weber had experience and interest in that area too:

> I was always very interested in the biological field from my time in Saudi Arabia before and during the first Gulf War and some of the work I had done in Libya at about that time. The opportunity arose because of the trust we had developed between our governments, because President Nazarbayev approved Project Sapphire. The opportunity that developed the next spring was to have a quiet visit to the world's largest anthrax factory in Stepnogorsk, Kazakhstan, just over the Russian border, as well as to a chemical facility and a biological weapons test center in Russia, so that was a full immersion for me into the Soviet massive offensive biological weapons program.
>
> What was so unique about it were the relationships we developed with the people through that process. It was my visit to Stepnogorsk. I had been briefed, I had seen satellite pictures, but going through that plant with people who could explain what it was capable of doing, it changed my life. It was an exposure to something evil that I really felt I should spend the balance of my career looking at, and it's a neglected area. Nuclear security, nuclear weapons, which I'm also passionate about, gets a lot of attention; the bio area has always been an orphan. It was a real niche that I could follow up on.[48]

And he did, and in the process, he oversaw a years-long effort to destroy the Stepnogorsk site, which is now a green field with no other sign of the global menace it once was. ("That certainly felt good," Weber told me.) But in the meantime, Weber had moved from the State Department's Foreign Service to the Department of Defense, where he joined then secretary of defense William Perry and future secretary of defense Ashton Carter to work on the broad Nunn-Lugar nonproliferation program that was housed in the Pentagon.

As for his deeper motivation, the ever-modest and soft-spoken Weber put it to me this way: it was "curiosity." "Look, it's one of those weird cases where you just do your job out of curiosity and you learn more about something and execute a project with incredible teams from the United States, from the Defense Department, from the Energy Department, and it's only twenty years later that you realize that it was something big that set a precedent for removal and consolidation of nuclear material around the globe. But at the time it was just a commonsense thing to do."[49]

A commonsense thing to do, indeed—through his efforts to prevent the proliferation of deadly nuclear and biological weapons of mass destruction, born of curiosity and fed by opportunity, supported by "incredible teams" from across the US government, Weber probably did more to save humanity from unspeakable horrors than all but a very few people in history. When we consider the ultimate impacts of US-FSU scientific cooperation, we cannot fail to take this amazing story into account.

Laura Holgate

Ambassador Laura Holgate's entry into the world of US-FSU scientific cooperation came about even more directly via the nonproliferation path. As a graduate student at the Kennedy School of Government at Harvard University, she worked with Ashton Carter and others on a report, *Soviet Nuclear Fission*,[50] that essentially laid out the scaffolding for what later became the Nunn-Lugar nonproliferation program, which Holgate then went on to manage for several years at the Pentagon. Later, Holgate became the principal Russian specialist at the Nuclear Threat Initiative, which directly funded several cooperative projects to reduce dangers from unprotected nuclear materials and was appointed by President Barack Obama as the US ambassador to the International Atomic Energy Agency (IAEA).

Rose Gottemoeller

Prior to her appointment as undersecretary of state for arms control and international security affairs in 2012 (and subsequently deputy secretary general of NATO), Rose Gottemoeller had a substantial résumé in global security affairs, including service at the Department of Energy as deputy undersecretary of defense nonproliferation, in which capacity, among other things, she oversaw the early stages of the Global Initiatives for Proliferation Prevention (GIPP) program at the National Security Council and at the Carnegie Moscow Center. For these and many other high-level policymakers, the choice to study and work in roles in which they could promote nuclear security and prevent global disaster was their entry ticket to the world of bilateral scientific cooperation.

While science cooperation was not necessarily or always their primary brief or preoccupation, these officials and diplomats had a major influence on the scope and direction of a major component of US-FSU science cooperation after the fall of the Soviet Union.

The Russian Language and Cultural Junkies

There were yet others who landed in the world of US-FSU science cooperation essentially by accident or circumstance. For these people, including myself, their long-term interest in the region came not necessarily through any particular interest in science, nor by virtue of a passion for global security, but from their study of, and in some cases addiction to, the language, culture, and history of the lands of the Russian Empire.

Norman Neureiter

We have seen how Norman Neureiter, one of the leading actors and thinkers on international scientific cooperation, became involved midcareer in this field. Reaching back further to his youth, however, Neureiter illuminated some deeper roots of affinity, through his mastery of the Russian language: "When I went to college, at age sixteen in 1948, my father, a college professor and an immigrant as an adult from Austria, said to me, 'Major in whatever you want, but also take Russian. It will be useful someday.' How perceptive he was!"[51]

It was his language competence and interest that drew Neureiter into the next phase: cultural exchanges.

But what made my Russian a lot better came in 1959. In a period of slight warming in US-Soviet relations, there had been an agreement between President Eisenhower and Nikita Khrushchev to exchange national exhibitions. The Russians came to New York and displayed their industrial capabilities with a variety of machinery. But the US decided to show the Russian people the material things of a typical American family. It was a huge event—forty days with fifty thousand visitors a day. Color television, automobiles, fashion, cars, a full-size model home with kitchen and bathroom appliances. It became the site of the famous Nixon-Khrushchev kitchen debate. In any case, I had applied and was selected to be one of the eighty Russian-speaking guides. Responding for eight to ten hours every day to the loaded queries of Soviet hecklers, as well as answering serious questions from Russians who wanted to know about life in America, can improve one's Russian fluency quite rapidly.[52]

Neureiter's professional language work on the cultural exhibits was followed by two more government requests to employ his skills, while he was working for Humble Oil in Texas as a research chemist:

Then in 1961 the government asked me to do something else. They asked me if I would serve as interpreter for a Soviet petroleum delegation which was coming to the US. The US oil industry had already completed their first organized visit to the Soviet Union and this was to be the reciprocal visit to the US. The US industry was very concerned as it appeared the Soviets were preparing to enter the global oil market for the first time.... Another invitation came in early 1962. I was asked to interpret for [the Soviet chemist A. V.] Topchiev. He was the official Soviet Pugwash delegate, but he was also vice president of the Soviet Academy of Sciences. He was an organic chemist and that was the reason that the US National Academy of Sciences asked me to do it. This was a very big deal for me.[53]

For Neureiter's employer, Humble Oil, however, it was the last straw:

This trip was a major event in my life. It was also the third time that Humble Oil had let me go off for several weeks on a diplomatic mission involving the Soviet Union. A few months later I was again offered an interpreting assignment by [the State Department], but it was clear that I could not keep taking time off from my job as chemist at Humble Oil but had to make a decision. Eventually I did decide to go to the National Science Foundation, then the Foreign Service, and then to the White House Office of Science and Technology, in all of which I personally held positions where science was an essential element of active diplomacy. Such activities led me to my current involvement in what we have come to call science diplomacy—the use of science as an active instrument of a constructive and peaceful foreign policy. It cannot solve all the problems, but I believe it can contribute to improved relationships and hopefully, over the long term, to a better and more peaceful world.[54]

Paul Hearn

Interest in Russian language, culture, and history were also critically important doors for myself and several colleagues who eventually led to careers in managing scientific cooperation with those and other countries. Of the four people whom I will present in the remaining part of this section, only one of us—Paul Hearn of the US Geological Survey (USGS)—had any formal science background, and even his was an afterthought to the siren call of Russian studies and Soviet studies that enchanted and lured us into the somewhat exotic field of US-Soviet science cooperation.

Hearn recalled:

> I had been admitted to Duke University. A big topic of discussion was which language you were going to choose because, at that time, it was required that you had two years of a language. . . . A friend of mine then said, "Take Russian, it's an easy B." . . . From a chemistry major . . . by the time I graduated, I graduated as a Russian major in 1971. . . . I was accepted into Georgetown for their Russian areas studies program, basically because I had no idea what I wanted to do and so it was essentially a fifth year of college for me.[55]

Still unsure of what he wanted to do, while at Georgetown Hearn took a part-time job at the Smithsonian Institution in a section studying minerals. They fascinated him, and he went on to get an MA in Geology from George Washington University and worked thereafter as a scientist:

> It was [also] a natural next step to have interest in the . . . nature of Russian geology, the actual geology of the Soviet Union as a physical part of the world, but also how they approached the science. . . . At that point in my life, I was totally ignorant of the structure of Russian or Soviet science, how it differed from us, and so any opportunity there was for me to go to a talk or if there were visiting scientists, I would jump at it. . . . I just naturally [initiated this] myself in terms of meeting Russian scientists and pursuing professional and personal relationships with them as friends and colleagues. That continued on further for the rest of my career.[56]

Hearn became the chief scientist at USGS in charge of cooperation with the Soviet Union under the intergovernmental Basic Sciences Agreement that was concluded in 1988. Paul and I spent many fun years together managing activities under that agreement and comparing notes about many things, from Soviet science to the wonders and agonies of the US interagency process.

Cathleen Campbell

Two other close colleagues, Cathleen Campbell and Gary Waxmonsky, also followed the Russian and Soviet studies path but, like myself, in more direct fashion, bypassing the detour in natural science altogether. Campbell's love of Slavic culture guided her toward the study of Russian as an undergraduate and on to receiving an MA in Russian and East European studies from Georgetown University, writing her master's thesis on Muslim populations in Central Asia. She worked for a time at a government-funded think tank, RAND, on policy issues, and moved into government positions at the Department of State and the White House Science Office, both focusing in large part on science cooperation with the FSU. Of her entry into work at the State Department on US-Soviet science and technology cooperation, she refreshingly and candidly shared with me a familiar refrain for humanities-based graduates of degree programs in the Slavic studies area (myself included): "I needed a job." "But who knew," she added, "that I was going to love the work so much?"

After serving for several years at the State Department and the Office of Science and Technology Policy, where she served as midwife to the renascence of US-Soviet science cooperation during the perestroika period, she joined the Commerce Department's Technology Administration with responsibility for programs across many countries. Having worked with her for several years, I recruited her to CRDF as senior vice president, and after a few years, she served as my successor at CRDF as president and CEO for ten years. Campbell, however, accomplished what I couldn't or wouldn't do at CRDF: making it a truly international organization by expanding its geographical scope to Africa and Asia.

Gary Waxmonsky

Gary Waxmonsky, who headed cooperative US-USSR programs at the Environmental Protection Agency for decades, was also a Russian-Soviet studies junkie, like Campbell and I. Waxmonsky's graduate school training was even closer to my own: he studied political science and Soviet political history under the tutelage of Robert C. Tucker and Stephen F. Cohen of Princeton University, who were also my advisors a few years earlier. Like Campbell and I, he wrote his PhD dissertation on an issue almost totally unrelated to science and technology: the life and times of Feliks

Dzerzhinsky, the creator of the *Cheka*, the Soviet secret police, which in later incarnations became the NKVD, KGB, and FSB. I suppose that his choice of dissertation topic had something to do with his shared Polish heritage with "Iron Felix." Indeed, both Campbell and Waxmonsky, like I and many others who entered the Russian Studies field, had some family background in the lands historically dominated in one way or another by the Russian Empire, and undoubtedly these roots had some role in our academic and career choices.

Gerson Sher

My own fascination with Russian language began very early, in a somewhat unusual way. My paternal grandparents were Jewish refugees from the tsarist epire in the first years of the twentieth century. It was very uncommon for Jews of that generation to use Russian in their homes; almost overwhelmingly, they wanted nothing more than to leave Russia behind them for good. It was not so for my grandmother, as it turned out. She had beautiful Russian-language editions of Tolstoy's *Anna Karenina* and *War and Peace* in places of honor on her reading bookcase.

I had heard of *War and Peace* (but not *Anna Karenina*) and filed away the thoughts, first, that my grandma was actually a very literate old Jewish lady who knew Russian, which I thought was a great oddity in itself, but also that there must be something very interesting about these Russian people that might be worth taking a look at if she brought these books all the way from Russia along with her Sabbath candlesticks.

I studied Russian in high school and once in college and having read *The Brothers Karamazov* in a world literature class, I was hooked. This was a very strange culture indeed: rationalism, wild emotion, and spirituality, all wound up into one messy bundle. I needed to know more. I became a Russian studies major at Yale, studied Soviet politics with Frederick Barghoorn,[57] read Nikolai Gogol' in the original Russian and became interested in dissident Yugoslav Marxist revisionism. I did not take a single science course. Not having the slightest idea of what I wanted to do next, like Paul Hearn, I enrolled in graduate school in the Princeton Politics Department, where I continued to study Soviet and East European political history and political philosophy, writing a doctoral dissertation on the Yugoslav Marxist group.[58] I "proctored" (taught undergraduate discussion sectors) for Steve Cohen's wildly popular lecture course on Soviet political history. Then it occurred to me one day that after all that preparation, I had no interest whatsoever in teaching. It was a crisis.

My salvation came from an unexpected quarter: science. Allen Kassof, who taught Soviet sociology at Princeton (and who had a wonderful collection of Soviet jokes), was also the longtime executive director of IREX (discussed in chapter 1). IREX had no openings, but he mentioned that the NAS in Washington, DC, had an opening and suggested that I talk to the staff director of its scientific exchange programs with the USSR and eastern Europe, Lawrence Mitchell. My initial reaction was something like, "Oh, no, anything but science!" But it was a job, and it was one in which I could use my language skills and perhaps my knowledge, though probably not my prodigious expertise on Yugoslav Marxism. I had a nice interview in DC, including a language drill, and I was hired.

In short, like Cathy Campbell, I needed a job, and I found one in managing scientific cooperation. It was a complete accident. Nothing had ever been farther from my mind, but for the next forty years, nothing seemed more natural. It was a front-row seat in the reality theater of Soviet and East European life—at least, that part of life that touched on the highly educated and privileged scientific community. It was also a way for me to continue to dabble in real-time intellectual history, though now the Soviet and East European scientific intelligentsia replaced my earlier obsession with the Marxist humanists.[59] But it was to reappear now and then as I tried to reflect on what I was doing and why.

The Entrepreneurs

It is important to bring into the story at this point a group of people who were generally not thought of as participants in scientific cooperation with the Soviet Union, at least not until 1991. At that time, the very notion of making money through access to the work of Soviet scientists was rather counterintuitive. Public opinion in the United States tended to regard the Soviets' development of the hydrogen bomb as little more than the result of espionage. Their successes in the space race—*Sputnik*, Yuri Gagarin, Laika,[60] and the rest—were regarded more as feats of engineering with ominous military implications than as scientific achievements. In the US government community, in particular, the accepted view was that the Soviets were "behind" the United States in every area of science, with the possible exceptions of physics and mathematics, and that as long as they continued to rely on "reverse engineering" and a state-controlled scientific establishment, they would remain hopelessly behind if we in the West were sufficiently vigilant and protective. The idea that they might actually have some competitive advantage in any science-based commodities or competitive technologies was not at the top of most observers' and policymakers' list of concerns.

John Kiser

But this line of thinking was not true for a small number of entrepreneurs who, like many innovators, dared to think differently. The first to debunk the notion of Soviet high-technology uncompetitiveness was a young man with an early interest in Russia, a business degree, and a talent for counterintuitive thinking, John Kiser. Kiser had visited the Soviet Union as a student, gone to business school, and decided to explore a unique business niche: Soviet technology. He sat down with a list of US patents to the Soviet Union from the 1960s and 1970s. As he reported in a ground-breaking article in *Foreign Policy Magazine* in 1976, "the rising number of patents annually granted the Soviet Union by the United States Patent Office . . . has jumped from 66 in 1966 to 492 in 1974. More significantly, in recent years there has been a growing number of Soviet technologies licensed directly to US firms, currently totaling twenty-five."[61]

He went on in the article to describe in some detail Soviet technologies in metallurgy, iron and steel, energy, transportation, and mining. Kiser made the following two important points, which might seem obvious today but in 1976 were quite unorthodox: first, that "the technology gap is largely a function of where one chooses to look" and, second, that "although Soviet manufacturing technology or quality controls may be backward by American standards . . . we should not conclude that the Soviets have nothing to contribute at the research level."[62]

The entire notion that there might be something to gain from Soviet technology ran head-on against the common wisdom of the day. A corollary, one that was constantly a thorn in the side of advocates of scientific cooperation with the Soviets, was that they had more to gain than we did and that indeed every interaction with the Soviets was a net loss to the United States. While uninformed, this view widely shared in the executive and legislative branches, making the arguments in favor of US-Soviet science cooperation a constant uphill battle.

Armed with his list of technologically advanced Soviet patents, Kiser obtained a research grant from the State Department resulting in a detailed report of an entire catalog of promising Soviet technologies and went on to found a business, Kiser Research, that for a fee assisted US companies to identify and obtain licenses to attractive technologies in their fields.[63] For the balance of the Soviet period, this was a successful business model for Kiser and his US clients. It was not the conventional business model, one in which an entrepreneur invests one's own money in a piece of intellectual property and then either sells it to others or builds a business herself based on the technology's market potential. But given the nature of the times and the oddities of the Soviet patenting and judicial system, Kiser's approach was sensible. After the USSR's collapse, however, Western companies could go directly to the source and were not in as great need for the type of facilitative services that Kiser offered.

David Bell

David Bell, president and CEO of Phygen Coatings Inc. of Minneapolis, Minnesota, had a different path to the world of Soviet science and technology. Like Kiser, he was neither a scientist nor an engineer; unlike Kiser, he was a technological entrepreneur who had worked for high-tech companies in materials, coatings, and photo imaging in the 1980s and 1990s before launching Phygen.

As I discovered late in my career, technological entrepreneurs are a special breed. They are people who are experienced in the art of combing advanced technology to meet identified market needs; people who, unlike ordinary mortals, take enormous financial and personal risks doing so, because it is never a sure thing; and people who understand, as it is often said in that community, that in order to be a successful technological entrepreneur, you have to have had three successive *failures*. Such people are born, not trained. Generally, scientists themselves have neither the temperament nor the interest in being entrepreneurs, though some do succeed, no doubt because both in science and in life they are risk-takers at heart. They have to be.

Bell recalled:

> I was part of a startup company that had acquired technology that was originated in Khar'kov, Ukraine. It was an innovation that would really revolutionize the world of thin-film hard-wear resistant coatings and materials because it was the development of titanium nitride coating in a concept or a technology

known in the former Soviet Union as *bulat* technology. The Russian manufacturing system had supply problems with getting cutting tools such as twist drills, end mills used in manufacturing.... The Institute of Cutting Tools in Moscow was a pioneer in developing and patenting exotic recipes for using thin, super-hard materials and depositing them as coatings on cutting tools to extend the life of those cutting tools and putting that system in manufacturing plants all across the former Soviet Union.[64]

Bell's task, however, was not to research or develop the technology, but to sell it: "I was hired as the marketing director ... and was given the task, 'How do you sell something that's never been sold before?' It was quite a learning exercise, but I ended up coming up with strategies to sell systems and technology to companies like Boeing, Chrysler, and General Motors.... The company was called Multi-Arc Vacuum System Inc., and it brought a new coating product to the marketplace, called 'tool recoating' that became the standard of reusing tools for making gears in the transmission manufacturing industry."[65]

Multi-Arc's business prospered on the basis of this revolutionary tool recoating technology. Subsequently, Bell bought the original company out and launched his own new small business, Phygen Coatings Inc. Today, Phygen is the leading producer of superhard coatings for automobile transmissions in the United States and is thriving.[66]

Ever the entrepreneur and looking out for new opportunities, Bell was also one of the first recipients of an industry-oriented US-FSU cooperative research award under CRDF's Next Steps to the Market Program. For many years thereafter, he worked with the International Center for Electron Beam Technology (ICEBT) in Kyiv, a spin-off of the Paton Electric Welding Institute, on physical vapor deposition of coatings using electron-beam technology. That project, while it led to very close personal and professional relationships, did not, however, result in the kind of commercial successes that Bell enjoyed from the much earlier *bulat* license.

By virtue of his CRDF grant, I became aware not only of Bell's efforts but also of the real possibility of developing concrete programs to catalyze commercially oriented scientific cooperation. This proved to be one of the most difficult challenges of my career but sometimes a very rewarding one. Bell was a key member of the board of directors of the US Industry Coalition (USIC), where I had my last full-time remunerated job, and he, among others, opened my eyes to the real world of small-business, high-tech innovation. No matter how many books you may read on the topic, and no matter how many "trainings" you may attend, you'll never understand it unless you meet and spend time with people like Dave Bell.

Randolph Guschl

By contrast, working for one of the world's largest multinational corporation, the DuPont Corporation, Randy Guschl had a scientific background. His initial awareness of Soviet science, however, came not through his work at DuPont, but by virtue of his assignment by DuPont to work at the Savannah River Laboratory of the US Department of Energy, which at the time was processing tritium and plutonium for nuclear weapons:

> I came out of two fine research universities, where I ended up with a PhD in chemistry from University of Illinois and ended up in DuPont. . . . As I worked for DuPont and rose through the management ranks and became an R&D director, we were encouraged to have exchanges, across borders and countries for customers and for science as a whole in areas that were rich where you had to have a cross-disciplinary approach. The Soviet Union was kept out.
>
> Then I took an assignment with the Savannah River Plant for the Department of Energy as a program manager, where . . . we considered the Soviet Union to be the enemy and we're supposed to be anticipating what they're doing . . . and what we've got to do. Suddenly, I started to develop a respect for what was going on there because I was privy now more to information about what they were doing. So that was where I first became aware of Soviet science, but we clearly were not going to be doing anything with it.[67]

It was only after Guschl returned to DuPont in the late 1980s that he was to be in a position to understand how Soviet technology might be of benefit to the company's commercial operations and to act on it. "In 1987," he recalled,

> I left Savannah River and came back to the commercial side of DuPont when DuPont left Savannah River. Three years later, I became an R&D director who was responsible for all of Europe, including the Soviet Union. But we had nothing going on in the Soviet Union except for a few deals through a manager I had in Germany by the name of Heinz Hefter. So we were aware of good science [in the Soviet Union] in proteolysis and other areas. . . . [Through] Heinz, we became aware of . . . which of the institutes [in the FSU] were really the high-quality thought leaders and which were protective of their ranks, if you will. We were interested.[68]

But interestingly, it was still not the direct corporate interests of DuPont that drew Guschl into the idea of scientific cooperation with the FSU but a government request. The Department of Energy was just starting up its nonproliferation program with the FSU, Initiatives for Proliferation Prevention. Guschl, in part by virtue of his prior work with the DOE at Savannah River, contributed to the DOE's efforts to conceptualize a nonproliferation program that would involve private industry, in accordance with its unusual congressional mandate: "So when I was asked by our government to look at some possibilities for some new programs—CRDF, [G]IPP, ultimately USIC and others—and went back to ask DuPont [for permission], people there said, 'Yes, we know there are good scientists there and there is good technology, but we don't know how to access it.' So I was asked by DuPont to explore how these programs might help our business needs."[69]

As it turned out, DuPont's greatest successes in working with the FSU science community were not in its traditional chemistry business but in agriculture. Pioneer Seed, the agricultural giant created in the 1920s by future US vice president Henry A. Wallace, was acquired by DuPont in the 1990s. Through programs such as CRDF Global and USIC, specialists from Pioneer were able to get access to seed libraries in Ukraine and Russia. These libraries were of substantial interest to a company like Pioneer, which provides finely customized seeds for various conditions to farmers in the United States and elsewhere. In this, Pioneer's experience was akin to that of the data-dependent scientists like Julie Brigham-Grette—earth scientists, botanists, zoologists, and others—discussed earlier in this chapter. In addition, Pioneer saw important market opportunities for its own products in Ukraine; this was an additional reason why it devoted substantial resources to working with that country.

Notes

1. Interview with Kip Thorne, October 6, 2015.
2. Ibid.
3. See *Review of U.S.-USSR Interacademy Exchanges and Relations 1977*, 90-92.
4. However, it is fair to note that formal intergovernmental agreements were essential enablers for all such activities. In particular, the 1958 Lacy-Zaroubin agreement, including its periodic renewals, provided the essential framework for all visits, formal and informal, since it created special visa categories and other kinds of provisions that gave them special protections and status not available to ordinary tourists.
5. See the first section of chapter 7, "Kip Thorne: Gravitational Wave Detection."
6. Interview with Tengiz Tsertsvadze, December 7, 2015.
7. Ibid.

8. Interview with Boris Shklovskii, January 14, 2016.
9. Ibid.
10. Ibid.
11. Sagdeev 1994, 71.
12. Ibid., 75.
13. Ibid., 76.
14. Interview with Marjorie Senechal, May 10, 2016.
15. Ibid.
16. See the section in chapter 11 titled, "International Women's Day (Marjorie Senechal)."
17. Senechal interview.
18. See the section in chapter 4 titled, "The Civilian Research and Development Foundation."
19. Interview with Lawrence Crum, January 6, 2016.
20. Ibid.
21. Ibid.
22. Interview with Yaroslav Yatskiv, December 15, 2015.
23. The Vega program involved the USSR, Austria, Bulgaria, Hungary, the German Democratic Republic, Poland, Czechoslovakia, France, and the Federal Republic of Germany.
24. Yatskiv interview.
25. From the late eighteenth to mid-nineteenth centuries, Russia colonized substantial parts of North America's Pacific northwest, in particular in Alaska and northern California. The Russian colony in California was at Fort Ross, near Bodega Bay, and lasted from 1812 to 1841. See "Outpost of an Empire."
26. Interview with Peter Raven, February 23, 2016.
27. Ibid.
28. Ibid.
29. Ibid.
30. See the section in chapter 7 titled, "Impact on Scientific Infrastructure."
31. Raven interview. When I visited Georgia in 2015 for this book, I heard about the flora of eastern Georgia again from Dr. Maya Alkakhatsi of the Institute of Botany; see chapter 5, "The Science."
32. Interview with Karl Western, July 15, 2016.
33. Interview with Julie Brigham-Grette, December 2, 2015.
34. Interview with Roald Hoffmann, February 18, 2016.
35. Marshall Shulman, one of the leading Sovietologists of the day, was for many years a member of the NAS Advisory Committee on the USSR and eastern Europe, which set the policy for the interacademy exchange programs. He was a very kind and gentle man and a great scholar. While I worked there, each year at the annual meeting of the Advisory Committee, he would deliver a talk about the current political situation in the Soviet Union. He always ended it by saying that it's a time of great transition—which of course was true, year after year. I doubt he could have imagined the immense transitions that occurred some twenty years later; no one could.
36. Hoffmann interview.
37. Ibid.
38. Interview with Loren Graham, October 19, 2015.
39. Interview with Siegfried Hecker, January 28, 2016.

40. Interview with Terry Lowe, March 26, 2016.
41. Ibid.
42. See Hecker, ed. 2016, 2:175–341.
43. Personal memorandum from Glenn Schweitzer, May 28, 2016.
44. See Hoffman 2009, 440.
45. Interview with Andrew Weber, April 25, 2016.
46. Ibid.
47. Ibid.
48. Ibid.
49. Ibid.
50. Carter et al. 1991.
51. Interview with Norman Neureiter, January 12, 2016.
52. Ibid.
53. Ibid. Neureiter's more detailed story of the Topchiev trip appears in chapter 6.
54. Ibid.
55. Interview with Paul Hearn, August 5, 2015.
56. Ibid.
57. Just a couple of years earlier, in October 1963, Barghoorn had famously ended up in the Lubyanka (KGB headquarters in Moscow) under arrest for espionage after foolishly distributing political attitude surveys on internal Aeroflot flights, occasioning the first use of the "hot line" by President Kennedy to obtain his release.
58. Published as Sher (1977).
59. I did, on the other hand, find time in the spring of 1979 to teach a full course load on Soviet domestic and foreign policy and on Marxist humanism as a visiting assistant professor at Duke University while working part time at the academy. It was a disaster.
60. Yuriy Gagarin was the first human to make a space flight, in April 1961. Laika, a dog, was the first animal to make a space flight, in November 1957, a month after *Sputnik*.
61. Kiser 1976, 136–37.
62. Ibid., 143, 145.
63. Kiser 1977.
64. Interview with David Bell, January 14, 2016.
65. Ibid.
66. Ibid.
67. Interview with Randolph Guschl, March 21, 2016.
68. Ibid.
69. Ibid.

6

WHAT KEPT THEM GOING?

We have seen how and why our various actors got involved in science cooperation between the two countries in the first place, but what sustained their continued involvement? Did their initial motivations remain the same as they developed their relationships, or did they change over time? In practice, both of these things occurred, but for different individuals, in different measures. What happened most often was that as the relationships deepened, the motives became mixed together in varying degrees.

The Science

It is fair to say that for all the scientists, those who remained engaged in the activity over the years did so at least in part because of the scientific interest it held for them, but the interpersonal relationships were also a major factor. For a great many, as they moved beyond the vodka-and-caviar stage, they formed more personal, deeper friendships and shared values. Kip Thorne, for example, while continuing to work on gravitational physics with his Russian colleagues, also developed such a degree of trust among them that he turned out to be the conduit for much of Andrey Sakharov's correspondence with American scientists and his public statements.[1]

This was an unusual case in terms of Sakharov's celebrity, but many scientists, and others, have stories of various kinds of the slightly more mundane risks they took to help friends they made over the years. I well recall, for example, how in 1988 or so, my colleague Paul Hearn agreed to take a serious assortment of artistic lacquered boxes out of the country for Russian friends who hoped to build up enough savings in the West to allow them eventually to emigrate in reasonable comfort. In fact, such stories are very common to any Western visitors who traveled to Russia more than casually. Coming back to science, another story I heard many times and that nobody has ever denied was that the first tranches of emergency assistance to Soviet scientists in the very early 1990s arrived literally in the form of cash concealed in the clothing of US national laboratory scientists from Livermore and Los Alamos.

Underlying these relationships of trust and empathy, for most of the scientists, was the science itself. There were exceptions, to be sure. Marjorie Senechal, during her year in Moscow under the interacademy exchange program, was surprised that there were fellow exchange visitors who did not share her paramount interest in the scientific significance of their stays: "People had their own motives for going there. You know, we saw all kinds of behavior among Americans there, and we were shocked at some of them. I wanted to ask one of them, who lived down the hall, 'Why the hell are you even pretending to be serious? You're just here for a lark.' But others were genuinely serious, and then [others were there] getting to know the people from other countries a little bit."[2]

In many fields, as we have already seen through the testimony of scientists like Thorne, Senechal, Larry Crum, Peter Raven, Rita Colwell, and many others, it was serious, high-quality science being done in Soviet or FSU institutes that kept their interest high. For others, like Brigham-Gretté, it was access to uniquely valuable geographic sites. The quality of Soviet geoscience was for her, at least at first, a secondary consideration. Moreover, since Soviet geoscientists tended to publish almost exclusively in Russian-language journals that were not available in the international literature, their work was often unknown and perhaps underestimated as well. Instead, what proved at least as important for American geoscientists (in a pattern we will see repeatedly in the "field sciences") was location, location, location—for example, Lake Baikal. As Paul Hearn of the US Geological Survey put it:

> I think the best products came from when there was a particular location, and Lake Baikal is a good example of this, where there was a particular geographic area that had geology that was unique.... Baikal's only analogs are the Great Rift Valley in East Africa in terms of a physical environment, but in the Soviet case, they had research experience and control over the area that had not been shared until things began to warm up in the eighties....
>
> Baikal is a rift zone. It's an area where two plates are slowly moving apart, but in kind of an oblique angle; they're not spreading with a ninety-degree vector, but it's more like maybe thirty degrees, so one of those plates is sidling off to the northeast ... and the other one is moving to the southwest. If you

look at the width of that lake, which is about sixty to seventy miles, that constitutes the amount that it's spread in the last twenty-five or twenty-six or so million years. So from the standpoint of an active rift zone, it is unique. Also, because of that, its entire ecology and all of the plants and animals and fish that are there are unique.... So it definitely attracted our very best scientists in terms of various aspects of studying that.[3]

But there was another important circumstance that explained the interest of US and other scientists in Soviet science; in particular, there was one that was not immediately obvious to those people in the West who tended to look down their noses at Soviet science. It was the very breadth of research in Soviet institutes, and specifically, the fact that basic and applied research were not as rigidly segmented as they sometimes are in the United States, despite the perception that either all Soviet scientific research was funded by the military, or that dual-use research was tightly compartmentalized in separate institutes or design bureaus of the branch industrial ministries.

The truth lay somewhere in between. It is generally accepted that some 70 percent of scientific research performed in civilian scientific institutions in Russia was funded by the military. However, in those same institutes, the best researchers enjoyed an unusual degree of freedom to choose their research topics and to delve into the most fundamental research on issues that may or may not have been directly related to the military needs of their funders. In part, this was one of the luxuries of the top-down, command-economy type of funding that characterized the Soviet science establishment in general. Funding was not project-specific; it was institute-specific, and the institute director had substantial latitude in determining which scientists and laboratories would be funded, and how.

Thomas Pickering's comments about his experience at Boeing give one a sense of the extent to which the boundaries between military and civilian research in the United States were in some ways more rigid than those in Russian institutes:

> One of the most interesting stories in the Boeing piece was that Boeing decided that the 787 would be a mainly composite airplane. And we wanted to use our Moscow design folks to work on pieces of that because they produced designs for us at a highly competitive labor rate and they were very good at it. Immediately, we ran up against the roadblock that since most all composite work in the US had been military, it was a Munitions Control licensing requirement to do that kind of work offshore with others.
>
> The president of our Russian operation, who worked jointly for me and for the commercial airplane leader, came to me and said,

> "How are we going to work with this?"
>
> I said, "Well, what do you guys know?"
>
> And he said, "Well, I know the guy who probably is the best guy in composites in Russia."
>
> "Well could you bring him to the States?" I asked.
>
> "Sure," he said.
>
> And so we arranged with the various contacts that Boeing had to have this guy come and spend a whole day with Defense, the intelligence communities, State Department people, and everybody else, and they asked him what he knew about composites. Toward the end of the day they said that he knew more about composites than they did. So we got a license. And we set up a compartmented facility in Moscow in a separate building. . . . He obviously put our guys in the position of saying, yes, this guy does know more than we do about composites, so we're not going to lose any technology. We're going to gain if the Russians start designing airplanes for us.
>
> But this was all on the civilian side, although within Boeing we can, if we don't break classification rules . . . migrate technology and processes between the military side and the civilian side as long as we defend an end game [about the final product].[4]

Thus, in many areas, not only American scientists but also the American commercial and military research communities were interested in what the Soviets were doing and, perhaps somewhat to the surprise of the more skeptical entrepreneurs and defense people, found that they could get access more easily than they had ever imagined, even though in the United States, similar research may have been carried out under defense contracts with strict project-based restrictions. This was in part because in the Soviet Union, the research was considered basic and was performed in nonmilitary Academy of Sciences institutes. Even if these institutes may have had restrictions on access by foreigners, it was possible to connect with their researchers, especially through the officially sanctioned exchange and cooperative research programs.

In 1982, at the height of aggressive actions by the US Department of Defense (DOD) to restrict Soviet and East European scientists from visiting the United States even for open scientific conferences, there was a fascinating exchange in *Science* magazine between then Executive Director of AAAS William Carey and Deputy Secretary of Defense Frank Carlucci. Carey had written to Carlucci,

> I must tell you that the otherwise excellent brochure on *Soviet Military Power* went off the rails badly . . . in contending that US-sponsored scientific exchanges and scientific communication practices enhance Soviet military power. I am dismayed to find the Department of Defense indicting

inter-Academy exchanges, student exchanges, scientific conferences and symposia, and the entire "professional and open literature" as inherently averse to US military security interests. . . . I find it deplorable to have our Defense Department taking a public and well-advertised stance that exchanges and the open scientific literature constitute still another window of vulnerability and a free asset handed to our principal adversary.[5]

Carlucci responded to Carey with a litany of complaints about Soviet snooping in areas of US critical technology, including electrometallurgy, in particular superplasticity and fracture mechanics. He proudly stated that "a concerned U.S. government scientist succeeded in stopping the exchange in these militarily related topics. However, it was dismaying later to find that the Soviets had acquired the information under the auspices of a new subtopic on corrosion."[6]

At the National Science Foundation (NSF), in fact, through my work in the Electrometallurgy and Materials Working Group, I was quite familiar with the DOD's concerns. I knew the "concerned US government scientist" in question and knew that he was in fact keenly interested in what the United States could find out about the cutting-edge work that the Soviets were doing on powder metallurgy, super-hard materials, and other fields certainly for their potential military application. At about the same time, Terry Lowe of Los Alamos was beginning to learn about the Soviets' work on nanomaterial coatings for military aircraft and going to international scientific meetings to learn more. None of this was surprising; that's what adversaries do. But the story does illuminate something that is counterintuitive: that in some areas, at least, where basic and applied research intersect, there may have been more holes in the Soviet science system than might have been supposed.[7]

What about the science-related motivations of the scientists from the former Soviet Union (FSU)? Again, there was a mix. Certainly, the familiar and often true (but often much too simplistic) perception of the United States as the world's leading scientific power played an enormous role. But it wasn't that simple. It is generally recognized that in certain areas, such as theoretical physics and mathematics, Soviet scientists were among the world's leaders. There were, however, "niche" institutes in other areas whose research was unique and ahead of its times.

One such institute is the Eliava Institute of Bacteriophages, Microbiology, and Virology in Tbilisi, Georgia, whose fascinating history merits a brief historical digression. Marina Tediashvili, head of the institute's

Microbial Ecology Laboratory, described the institute's origins being born of international scientific cooperation in the 1920s: "George Eliava was a physician, a medical doctor and was also educated abroad. He met [Felix] d'Herelle [the codiscoverer of bacteriophages] . . . at the Pasteur Institute in Paris, where he worked for several years in the 1920s. . . . D'Herelle also worked here in this institute. . . . Then, the laboratory in Tbilisi was converted into a public health lab, a bacteriology lab, and then on the basis of that lab, the Institute of Bacteriology and later the institute of Bacteriophages was created. Eliava was the first director."

In the 1930s, during the height of Stalin's purges, Eliava ran afoul of Lavrentii Beria, at the time the secretary of the Communist Party of Georgia. Tediashvili continued:

> But then, George Eliava was executed in 1937 . . . as an enemy of the people. . . . They say that d'Herelle sent a letter to Stalin or called or somehow contacted him, asking to help with George Eliava, and they say that Stalin called here, to Beria—Beria was at that time the top person here, in Georgia—asking about George Eliava. Beria told him that he's already dead. . . .
>
> There [was] a book published in 2004 in Georgian written by Avandia Khichuri. a professor of medicine, about the medical doctors and medical researchers executed in those horrible years. . . . He found that there were a number of "official" reasons for Eliava's arrest: that he was claimed to be a French and British spy, and that he was developing plans [on] how to poison his enemies (Beria among them), that he was poisoning people with the bacterial cultures (like plague preparations), killing children, something like that. . . . But what we do know is that in this institute . . . a great many aristocratic family members were working here you see (some of them as researchers, but mainly as support personnel, technicians). [Because] the aristocrats were suppressed, [Eliava] was trying to give them work here.[8]

It would also seem that another, perhaps more powerful reason for Eliava's demise was a change in the nature of the institute itself. It began working on biological warfare, in particular anthrax.[9] This, as Tediashvili explained, had a profound effect on the institute and its international exposure.

> In the Soviet period, the Eliava Institute was a type of semi-closed institution . . . and so publications were not the main priority at the time. Of course, we had made reports . . . but beginning in the '90s, of course, we wanted to have more publications. Until now, perhaps one [of] our main weaknesses . . . is that a major part of our previous work and experience had not been published. Now, we're working on preparing some big reviews, or essays in books on what we had already completed. One of these books was funded through the UK Ministry of Defense.

Today, once again, international cooperation is vital to the Eliava Institute and the vitality and recognition of its unusual scientific work on bacteriophages:

> [On the problem of publication,] I can tell you for sure that without international cooperation and international relationship, we would not be able to publish in the international journals. Even now it's quite difficult to do it because, despite having quite well-equipped labs. again mainly due to international grants, we are still behind the Western research centers in terms of our technical/instrumental and of course financial capabilities.... So we need collaboration. Our publications now are based on the valuable experience of Eliava's scientists, especially in the field of phage research and phage therapy, just a kind of mix of our [past] experience, really, and [more recent] modern investigations, [in which we can use] the best laboratory resources.... And of course, all this is the [international] collaborative work of the scientists. We are writing these manuscripts together.[10]

Much of the credit for "rediscovering" the Eliava Institute after the fall of Communism belongs to Dr. Elizabeth Kutter of Evergreen State College in Olympia, Washington. Kutter's long collaboration with the Eliava began with a four-month research visit in 1990 under the interacademy exchange program at the Soviet Academy of Sciences' Engelhardt Institute of Molecular Biology in Moscow to research bacteriophage genetics. While there, she made contact with the Eliava team and thereafter pioneered and cultivated, both through research grants and personal funds, strong scientific ties between the Eliava and the world scientific community.[11]

International cooperation, especially since the breakup of the Soviet Union, is today a lifeline for many ex-Soviet institutes. In Georgia, Maia Alkakhatsi of the Institute of Botany in Tbilisi, made this clear in her interview but put more emphasis on access to modern research equipment than to scientific publications:

> When we work with other people, they have different approaches.... We provide much information [on] what we have here and how it grows.... And then there are possibilities that [foreign visitors] will finance something.... If we have a project, then there are computers and that they come through projects.... Here we have many computers and microscopes [that] come from [cooperative] projects.... If we just have contacts and we work together, that leads to publications in international journals.[12]

Access to modern research equipment, especially, was a major magnet for FSU scientists at all times. During the Soviet period, such opportunities existed only by spending time in foreign laboratories. US scientists and government agencies were not in the habit of sending equipment to Soviet institutes; it was extremely difficult if not impossible. After 1991, with the revolutionary change of paradigm from cooperation between adversaries to assistance of a desperate FSU science community, this changed dramatically. At Soros's International Science Foundation (ISF) and later at the Civilian Research and Development Foundation (CRDF), when we made competitive research grants, about half of the funds for the FSU side were devoted to procuring equipment and instrumentation from the West. Everyone understood that while salaries and travel were short-term benefits, hardware was of lasting value, and if you wanted your programs to have any long-term impact on science, purchasing capital equipment was on the critical path.

Alkakhatsi also commented on American and European scientists' interest in working in Georgia. It is useful in reading this to recall Raven's comments about the Soviet Union's ecological diversity and richness and his exposure in particular to the eastern Georgia biome:

> Foreign scientists have a lot of interest in Georgia and what we have here: good cultivated crops. For them, it's an opportunity. For us, of course, there's the opportunity for scientific contacts with foreign scientists, who give us the possibility of publishing in rated journals to study material on equipment that unfortunately we do not have.
>
> In East Georgia [where the Lagodekhi Reserve mentioned by Peter Raven is located], there are several areas we know of where we have many relict species, shrubs, others. . . . [The Americans] are interested in using the species for comparison for use as foods as corrective plants because it's very different. They are interested in what we have here, and they can use this as a resource. We are now a genetic resource . . . . People come here from America and they have identified which species they need, which seeds they need, and . . . in genetically modified crops they are using several genes from our collection.[13]

Returning to the motivations of scientists from the FSU in scientific cooperation, Paul Hearn, who had a great deal of personal contact with Russian scientists both as a scientist and as an administrator, discussed his perception of what kept them engaged. He told me that it was a blend of reasons:

> I think there was an interest, an honest, sincere motivation to support and forward their research by seeing the work of other colleagues who, up to that point, had not been able to see or had only read their papers and to meet their colleagues. . . . So, just from a purely academic standpoint, I think there was, very clearly, a keen interest in beginning that process of talking to their

counterparts, learning more about how they lived and beginning to understand them and know them not only as colleagues, but as people and becoming friends. But there was also a motivation . . . of [curiosity]; it was almost as if you're traveling to another planet and everything, from soups to nuts, is so different than what you have been accustomed to for your life up to that point. In Russian it would be called *egzotica*—exotica.

And that was not all. Alexander Ruzmaikin, who had been a research associate at the Soviet Academy's Institute of Applied Physics, was able to collaborate with a group of scientists in Great Britain. It was clear to Ruzmaikin that his work was on a par with their own, even if some of them did not want to accept that. (At an international conference, he met the British physicist Michael Proctor, who asked Ruzmaikin if he knew that his group at Cambridge was working on a certain problem. Ruzmaikin told Proctor that one of his students had already solved the problem for his PhD thesis. Proctor, according to Ruzmaikin, was shocked and upset.)

But Ruzmaikin did develop a good working relationship with two other British scientists, Andrew Sauer and Paul Roberts, who visited him in Moscow. When I asked Ruzmaikin, however, how important the joint work was to him scientifically, his response surprised me: "[It was] mostly important because of learning English," Ruzmaikin told me, "not in the sense of language but in the sense also of writing and presentation. Because the Russian language is implicative, if you understand what I'm saying. They write something assuming that you understand what they mean."[14]

This observation fascinated me, because I had already been tossing around thoughts about science and language, as it relates to the interaction between science and culture in the host country. I followed up on the language question in other interviews, and I will return to these thoughts in chapter 10, when considering the perspectives that my discussants offered about science in the Soviet Union and Russia.

Ruzmaikin's interest in using his collaboration with British scientists to improve his professional language skills also speaks to a much broader and deeper motivation of scientists from the FSU and, indeed, from all over the world: to be part of the world scientific community. This was especially important for Soviet scientists, and not only because of their deep isolation during Soviet times. They had a strong desire to be part of world science and they knew that, absent those limitations, they had a decent chance of being able to publish in the forbidden international scientific literature.

The desire to become part of the world scientific community was, in my mind, without a doubt the single most powerful aspiration and motivation of all Soviet scientists of quality to engage in any way they could in international contacts and cooperation. Of course, this is a deep cultural phenomenon, too, because the culture of that community—its values, its practices—differed very substantially from the culture of their home country.

For me, additionally, this fundamental tension raises another important question: "Does science really know no borders?" The common wisdom in the scientific community is that it does not, and a great many scientists in all countries subscribe to the notion that science knows no borders as a fundamental truth. Based on my forty years in managing international scientific cooperation, I have a somewhat different take: It is not a truth but an article of faith. In fact, science does know borders, but it doesn't like them.

Promoting Foreign Policy

A political goal—to improve the understanding and perception of the United States among the government and public in other countries, including our adversaries—was basically responsible for putting in place the very instruments that made cooperation between US and Soviet scientists possible in any meaningful sense. This was the philosophy undergirding President Eisenhower's people-to-people exchange programs of the late 1950s, of which the scientific exchange programs were essentially a special case. With time, Secretary of State Henry Kissinger added a second overlay, one relating much more directly to the implementation of foreign policy, to scientific cooperation: to serve as an incentive for restraint in the behavior of our adversaries.[15] Kissinger's approach was at once both more ambitious and more problematic, because it introduced at least implicitly a metric for evaluating outcomes: the more restrained the adversary's behavior, the more successful were the cooperative science programs in their fundamental policy purpose.

Ambassador Thomas Pickering is one of the most prominent proponents of, and practitioners of, science as an instrument of foreign policy. In my interview with him, he offered a more balanced and nuanced understanding of the importance of science and technology cooperation for foreign policy: "I think it's very important—science as an instrument of foreign policy, and foreign policy as an instrument of science. And what it does is to build a basis for trust and understanding if it is handled in a transparent, logical, and scientific way. If it is not handled as an instrument of raw propaganda solely. . . . It enjoys a special position internationally because of our accomplishments in science. But we're not unique in that regard and the Russians certainly over the years have made their own major contributions."[16]

Pickering had a deep appreciation of the relationship between science and foreign policy as a leading practitioner. In the 1970s, he had served as assistant secretary of state for oceans and international environmental and scientific affairs, and immediately prior to his appearance in Moscow in May 1993, he served as ambassador to India. Of his experience in India, he told me,

> As the USSR began to disappear, the Indians had to pretty fundamentally reexamine the relationship with us, and it was the science relationship that had kept things going for those years. . . . One of the things that has made a huge difference in India is that early on, four or five countries decided that each one of them would build a world-class scientific and technological institute in India and work with their best universities to develop faculty and the curriculum. We did one in Kanpur, and others—Germans, French, British, and I think even the Russians—did one. And that is now the IIT, the Indian Institute of Technology. And that fed into things like Silicon Valley, which we think is a totally American conception, a totally American opportunity. And so you talk about the internationalization of science—how can you possibly exclude the Indian contribution? Nobody stands on their own shoulders.[17]

We can get useful insights into the nature of the science/foreign policy relationship in the minds of diplomats by following the chronology that Pickering used to characterize its origins and evolution in US policy over the past sixty years. The earliest years of bilateral scientific cooperation, the late 1950s, corresponded with Pickering's early years as a Foreign Service Officer. In those years, he said, "It was really how in what way science could help us best to inform ourselves [the United States] and then be used as a point for finding agreement on some of the touchy issues." One of those touchy issues, within a few short years, was the Cuban Missile Crisis as well as its aftermath:

> Initially, a huge breakthrough was that both sides realized, particularly after the 1962 Cuba problem, that they had to be extremely careful about how they both built and structured and handled and controlled the deterrents [the nuclear weapons]. And that being able to in fact strengthen what one would call the structuring and indeed the knowledge of each other—the people who had some influence over management of the deterrent—were philosophical

ideas but they had very strong practical application. And so that was helpful. And all of this was *informed* by science and in some ways, obviously, stimulated by science.[18]

"The second period," Pickering continued, "was really how can science programs begin to strengthen and build détente, much of which stemmed, interestingly enough, from the common interest in arms control." These are what I called in chapter 2 "the détente years." In this period, Pickering noted, "we were doing things like magnetohydrodynamics; we were doing a bunch of energy research—things that we both thought were very interesting and important."

Pickering was ambassador to Russia from May 1993 to November 1996. These were the turbulent years of almost complete economic collapse, including the collapse of the scientific community. Pickering, as well as his deputy chief of mission, James Collins, who himself later became ambassador to Russia, were both extremely supportive of all efforts to promote scientific cooperation in those years. When I worked for George Soros from 1993 to 1995, I checked in with Collins several times during my countless visits in Moscow. Although Soros's ISF was a completely private effort, the embassy was deeply interested in these efforts. High-level professional diplomats like Pickering and Collins understood very well the importance of ensuring that the best scientists in Russia did not leave either Russia or science for a wide range of reasons—including the possibility of a massive "brain drain," the implications for global security, and not least of all the special role of the scientific intelligentsia in Russia in the promotion and defense of progressive values and human rights. I believe that without the "virtual roof" provided by such understanding and supportive individuals in government, the ISF—which had no relationship whatsoever to any official, intergovernmental cooperative agreement in science and technology—could not have succeeded in bringing over $100 million of emergency support to scientists throughout the FSU during these years.

Pickering describes the following period—the years after 1996—as one in which the policy goal was to determine "how and in what way can we reap commercial and, in the long run, foreign policy advantages in having a close science relationship."[19] From my own vantage point, however, it was not at all clear that significant commercial benefit from science and technology cooperation was a realistic goal, much less an expectation. However, some large American companies did enter the arena at the time. One of them was Boeing. Pickering, who, following his stint as ambassador to

Russia and as undersecretary of state for political affairs, became Boeing's vice president for international affairs, talked with me about Boeing's work in Russia: "We did a huge amount at Boeing. Beginning in 1991, Boeing had developed contracts which still exist with about twelve institutes of the academy in Russia to do advanced aeronautical research for them. And in '97, Boeing started an airplane design operation in Moscow, which now has fifteen hundred people working in it. There was a strong buildup then of globalization and practical application of science, but the Russians did stuff for us at Boeing that was extremely useful and continues to be."[20]

Another eloquent, distinguished, and energetic spokesman for a close relationship between science and foreign policy is Norman Neureiter, whose journey into science and technology cooperation we followed in chapter 5. An event that deeply influenced Neureiter's lifelong commitment to this field occurred in 1961, with the visit of the Soviet organic chemist and Academy of Sciences vice president Aleksandr Topchiev to the United States. The event was billed as a scientific tour of major US universities, but it had a much deeper significance. Neureiter, who was working as an organic chemist at Humble Oil, was asked by the State Department to take leave and serve as the interpreter:

> For me, this was a very big deal. . . . [Topchiev's] organic chemistry lectures, truthfully, were pretty modest in their content and conclusions. I translated the speeches into English. . . . But the real story was far more than that. In Washington, DC, at the National Academy of Sciences, he gave a talk. When the talk was over and most of the audience was gone, the Russians and some of the Americans went into another room for the reception. It was there that I found out that the real purpose of the mission was to talk about a Nuclear Test Ban Treaty. . . . It was an amazing trip for me and my first experience in this aspect of Soviet-US relations.
>
> We came to Washington just at the time of the 1961 annual meeting of the National Academy of Sciences. President Kennedy had been invited to speak to the academy members and our delegation was also invited to attend. We were seated in the front row and after making some brief remarks, President Kennedy left the podium, came over and shook hands with each of us. He spoke briefly with Topchiev, ending with something like, "Maybe you can tell us how to do it," referring to some recent success the Russians had had in their space program.[21]

Interpreting for Topchiev's meetings with American scientists, despite Neureiter's fluency in Russian, was an interesting challenge:

> On that same trip, Jerome Wiesner, then President Kennedy's science advisor, invited us to a dinner in his home, which turned into a very serious discussion about the test ban treaty issue. It was at that dinner that I began to reflect on my

role of interpreter. Two people who know little or nothing about each other are having a deep discussion about extremely delicate and serious issues. Neither understands a single word of what the other is saying—only the words that the interpreter chooses to use in the hope that the real meaning of what was said is communicated by those words. I later learned of famous mistakes that have complicated high-level diplomatic negotiations, but I worked very hard to convey the precise meaning in what was being said in both directions.... To the speakers, the interpreter is almost taken for granted and hardly noticed, but I came away with the deepest respect for the professionals who do this for a living and a realization of how critically important they truly are.[22]

The Topchiev trip, Neureiter recalls, "was a major event in my life." Among other things, Humble Oil told him that he had to make a decision to work for Humble or to go off on interpreting gigs for the State Department. After one more such trip, he left Humble and joined the NSF. From that point on, Neureiter was sold on science and technology in support of foreign policy, and vice versa. "But look," he said in his interview,

I'm really fanatic about the importance of engaging with the enemy.... I mean using science as an instrument of active engagement with bad countries or with countries we define as bad, with whom our political relations are really bad, but where a dialogue of some kind, a constructive dialogue of some kind, can take place. I happen to believe in engagement with the enemy. And you can argue against it, but unless you can starve them to death or something so they come and beg for surrender, which they don't do anymore if they ever did, nothing else works. So I feel very strongly about this.[23]

There were other, slightly more indirect foreign policy benefits from science and technology cooperation, and these were highly valued in the diplomatic service. Glenn Schweitzer recalled:

When assigned as the science officer at the US embassy in Moscow [in 1962], I had the opportunity to visit the academies of sciences and associated institutions in each of the fifteen republics of the USSR, of course including the academy in Moscow which did double duty as the academy for Russia and the academy for the USSR. While embassy officials were skeptical of the payoff from such travels, they soon became strong supporters after hearing my reports of the excellent receptions that I received everywhere. The academies were truly flattered to be visited by someone that they considered a representative of American science, and they welcomed the international interest in their efforts. Most important, I was usually able to program a visiting American scientist or a group of scientists to follow in my footsteps and begin serious conversations about specific aspects of S&T that were of mutual interest. Meanwhile, the political and economic officers who also visited the capital cities of the fifteen republics remained frustrated by the lack of access and interest they encountered within and beyond Moscow.[24]

The person with the science and technology portfolio, therefore, was in a way the ticket for other embassy officers, including highly placed ones, to obtain access to Soviet officials and citizens that they could not achieve on their own:

> The Cultural Affairs section of the embassy was a very useful partner as we knocked on adjacent doors of the Soviet government and institutions throughout the country. They often invited me to accompany them on visits to institutions of common interest—libraries, historical societies, museums, universities—and we were sure to include them in our receptions and other activities where the "culture" as well as the politics of the USSR was on display. Also, at meetings involving exchanges, and particularly student exchanges, we shared one side of the table.
>
> The counselor for Cultural Affairs had an apartment across the hall from mine, and when either of us had a reception, dozens of Russians would show up. When the counselor for Political Affairs or for Economic Affairs had a reception in nearby apartments, Russian guests were far outnumbered by diplomats from other countries. Indeed, if more than one or two Russians showed up, it was a success story. And when the ambassador had receptions for scientists at Spaso House [the US ambassador's residence], he counted on me to ensure that Russian scientists were present so that the ratio of Russians to others would be respectable. I was usually able to oblige and ensure that one or more cosmonauts were among the first to arrive, with many other scientists also in attendance.[25]

Global Security

After 1991, a new layer of deeply held concerns colored the motivations of US officials and scientists alike: the possibility that weapons and technologies of mass destruction, as well as their creators in the scientific and technical communities, could flow to other countries due to lax security and general chaos at home. The main goal of the science engagement programs, according to Rose Gottemoeller—who in the 1990s had been the National Security Council's chief advisor on Russia and the newly independent states and then went on to lead the Department of Energy's nuclear nonproliferation programs in the late 1990s—was to "wrap Russian and Ukrainian scientists in a web" so that "they would not go elsewhere." In addition, she said, it was hoped that these relationships would also help to "build the foundations for democracy," adding that it was a case of "American optimism."[26]

Laura Holgate was one of the chief architects of what came to be known as the Nunn-Lugar nonproliferation program and was its first director in the DOD. She told me that "the work with former Soviet WMD [weapon of mass destruction] scientists under Nunn-Lugar was seen as important for two basic reasons: first, there was concern that they might contribute to improvements in the Russian nuclear arsenal; and second, fear that they could contribute to the WMD capabilities of rogue states, such as North Korea."[27]

Holgate was careful to note, however, that at that time, the policy concern was about the proliferation of WMDs and their technology to states, such as North Korea, and not to nonstate actors. "The latter issue simply had not crystallized at that early date," she said; the possibility of nonstate actors or terrorist groups was simply not on the policymakers' radar at the time.[28]

Sig Hecker, the director of Los Alamos National Laboratory, formulated the global security issues that were preoccupying policymakers in terms of four worries: "If you look at the American concerns in the 1991–92 timeframe, [there were] four things that we worried about: loose nukes, that's the weapons; loose materials, plutonium and highly enriched uranium; loose people; and loose exports, 'sell everything that you can'. So those are the four problems."[29]

In addition to those concerns, the lab scientists had somewhat different concerns of their own. They looked at the situation from the standpoint of the Russian scientists, as well as that of the US government:

> So that was the American government's concern. The lab people had a sort of different concern. David Hoffman, a Washington, DC, journalist, wrote the book *The Dead Hand*.[30] In *The Dead Hand*, he summarizes what Nunn's and Lugar's concerns were. He says, "Those guys in Russia after the breakup of the Soviet Union, they were stuck with an inheritance from hell." That's what he called it, an inheritance from hell. Everything that you looked at, when Americans looked at the Russian nuclear complex, was the perfect nuclear storm that was about to happen. Those four things.
>
> When we looked at it, when we were over there, we saw the people. To them, their nuclear complex was a gift from heaven. It was not the inheritance from hell. It was going to save them. It was going to save Russia. So they had these dramatically different ideas than the governments of the two countries. We, the lab people, we understood, we knew they had to take care of their weapons. We knew they had to save them. [Our government] went after them to get rid of the weapons; we just felt that whatever the weapons the politicians decided they were going to have, by God that better be safe and secure.[31]

It was not only the Russian nuclear scientists and their weapons that were of concern. Marina Tediashvili worked at the Eliava Institute in Tbilisi, Georgia, where in addition to relatively benign research on bacteriophages, they also worked with extremely dangerous pathogens that were also potentially powerful WMDs:

> Also, here at the institute, we had been working on anthrax, Brucella, and clostridial pathogens—I mean, developing preparations for diagnostics, prevention, and treatment of this dangerous disease. . . . The anthrax vaccine STI was developed and produced here for the first time. Also at the Eliava Institute, the antianthrax immunoglobulin was developed and produced. [It is] . . . a very important preparation globally that survives to this day that has helped to rescue many people. It was used in the Sverdlovsk incident [the accidental release of anthrax spores from a Soviet military facility in 1979], because the antianthrax immunoglobulin was at that time one of the most important preparations in the world for treatment of acute, severe forms of anthrax. . . . Many other antibacterial and antiviral preparations were made here, so we were a major manufacturer of bacteriophage preparations and remain so today, as well as some different preparations, not only phage.[32]

The US government took note of the presence of anthrax at the Eliava Institute, which was a relatively open scientific facility administering phage and other therapy to the public, and designed a special program to address these concerns. The DOD, through the Defense Threat Reduction Agency (DTRA), funded and oversaw the construction of a separate, secure facility to store the anthrax spores and other pathogens in Tbilisi. Tediashvili explained:

> One of the reasons—aside from our huge experience in developing and producing phage preparations—why they became interested in this institute was especially because of the dangerous pathogens on which we had been working. And we had some collections of bacteria, but . . . now this collection of special and dangerous pathogens, our collection, is kept at the NCDC, the National Center of Disease Control . . . in Georgia. . . . The DOD created it and now there is a new Lugar Center, which is a reference lab under the Georgian Ministry of Health. Initially it was under the Ministry of Defense, but now they passed it to the Ministry of Health as a public health authority. . . . So now, then, there is a repository for the strains for special dangerous pathogens.[33]

In addition, they provided liberal funding to the Eliava Institute itself to upgrade its facilities, allowing it to keep its scientists gainfully employed in civilian research. This DTRA program continues to this day, one of the very few remaining US government–funded nonproliferation programs to survive and operate for so long.

Anthrax, as we have seen, was also Andrew Weber's preoccupation in the enormous project to destroy "the world's largest anthrax factory" at Stepnogorsk, Kazakhstan: "We learned quite a bit about rare diseases like Ebola and Marburg, on which they had big expertise, and we applied it for medical countermeasures and research. It was very, very important work.... It was joint research between the Vektor Lab, [Lev] Sandakhchiev's lab, and the US PIs [principal investigators] generally came either from the Centers for Disease Control and Prevention.... [or] the biodefense labs, because there were only a couple of labs that work on these rare diseases in the United States."[34]

World Peace

This story brings us to what was surely the most widely shared motivation for engaging in US-FSU science cooperation: promoting world peace. While only a few of my discussants specifically mentioned this wish, there can be no doubt that it was universally shared at one time or another, and to some degree or another, by every single person who was ever involved in this historic undertaking.

Larry Crum put it succinctly this way: "I think it was [Sen. J. William] Fulbright who said that if Reagan would have spent a Fulbright scholarship and gone to Europe and spent some time outside the United States, rather than all his life in Hollywood, he would have had a whole different impression of the FSU. And I feel the same way."[35]

Others expressed similar feelings. For Loren Graham, there was a

> hope or aspiration ... that by making contact with the Russians and by working with other people who wanted to make contact with Russian scientists, we would create a better world and that Russia might become more fully a member of the Western community. I thought science was the best bridge for this, better than, say, politics or literature. Music might have been another path, but I didn't have that kind of talent. But I thought science was a good bridge. I mean, science is supposed to be international, right? So I thought that a better world would be created.[36]

Peter Raven told me, "I've always been a dedicated internationalist. I believe in bringing nations together to promote peace, but even more fundamentally than that, I believe that it is interactions between people that allow each of us to accomplish the very best of which we are capable. If people talk to one another and come to understand one another, everything becomes easier, and the purpose of our life on earth becomes clear. I've always been a major promoter of such interactions, and always will be."[37]

For me, too, from my earliest engagement in Russian studies, the idea of somehow using this knowledge to promote world peace was a powerful source of motivation and remained an ever-present influence that kept me going even when things looked tough. Looking back on it, I do wonder what specifically it accomplished in terms of building a more peaceful world. I have no doubt, however, that the broad network of strong personal and professional relationships that the entire range of cooperative science programs made possible was one of their most lasting achievements. These personal relationships were by far the most commonly cited lasting outcome by my interview sample, from all countries. It comes down to this: When you can think of a colleague or friend in the FSU in personal terms, it becomes a lot harder to make sweeping categorical statements and rash judgments based on them.

Promoting Democracy

For many people, promoting world peace is synonymous with promoting democracy and democratic forms of government. For political scientists, though, there are plenty of troubling questions about this notion. When we talk about democracy, are we thinking of political institutions that are—or look—democratic? Are we talking about the principles articulated in written or virtual constitutions? Or about political and social norms and behaviors? And what about relatively stable democracies, such as the United States, as compared with unstable democracies, such as Weimar Germany?

Nevertheless, at some deep level, the view or sentiment has been widespread among many American, Soviet, and post-Soviet citizens that democratic forms of government—and values such as freedom of speech, association, worship, and due process—are desirable and will in the long run promote peace between nations (Lord Keynes's famous dictum notwithstanding). That was certainly one of the ideas that motivated me and others to get involved in Russian studies in the 1960s in the first place. While President Eisenhower did not, as far as I know, actually say that people getting to know people would promote democracy, the idea was certainly implicit that these contacts would lead to greater openness and awareness of the way other countries conduct themselves and the hope that some of it would rub off in the process. Soviet citizens, in particular, were immensely curious about the functioning of the American system, and Soviet scientists who traveled to the United States had the rare chance to see it in action, warts and all. This was a highly prized perk of the exchange and cooperative programs.

The association of democracy and world peace, I would suggest, is a very deep one. If not always a reality, it is a hope shared by many, except perhaps the most hard-core practitioners of realpolitik. George Soros's dedication of effort and massive funding to the goal of creating "open societies," based on the political philosophy of Karl Popper, is based fundamentally on the aspiration of creating and sustaining democratic values and behaviors throughout the world. I found much resonance with Soros's views and his passion, especially in the early 1990s, when everything seemed possible (at least, to my naive way of thinking). When I was working at the ISF with the leverage of a huge amount of money to introduce NSF-style peer review in Russia, the thought that promoting merit-based, bottom-up funding practices in the scientific community would be a tangible step to reinforcing norms of impartiality, due process, and the rule of law was certainly on my mind, though I never imagined that the adoption of peer review would pave the way for political democracy. I am certain that I was not alone in this hope, that it was in fact shared not only by my American colleagues but also at least to some extent, with a healthy dose of skepticism, by my Russian friends as well.

What have our actors had to say about it? Only a few actually did so directly. Loren Graham, who has thought and written deeply about US-FSU scientific cooperation as well as the practices and attitudes of Soviet and Russian scientists, put it in the following, very nuanced way, talking about our shared experience in the CRDF's Basic Research and Higher Education (BRHE) program in Russia to strengthen research and education in Russian universities:

> Let me address the question of whether we were doing this out of a spirit of generosity and help to Russia, or whether we were doing it out of our own self-interest. The happiest of all situations is when the two coincide: when what you're doing what you think is generous and good is also in your own interest. And I think that that's what was going on in BRHE. Look, if Russia would become more like the other countries of Europe and in particular, more like the United States and its educational and scientific system, that's both a good thing, in my opinion, and also in the interest in the United States. When we visited in 2005 with the American ambassador William Burns and other people and described what we were doing, the ambassador said, "We believe that anything that helps Russia to become more like another Western country is good; therefore, we approve of what you're doing." We didn't do it because the State Department told us to do it, but what we did, the State Department decided was a good thing.

But let me back up for a minute. Are democracy and openness American ideas? No. If you would tell a Frenchman that democracy and openness are American ideas, he'd tell you pretty quickly where to get off. He'd say, "We kind of had those ideas before you did, you borrowed from us." Or he might say, "We don't care where they came from, but they're as much French as they are American."

So when sometimes I was in Russia for the BRHE program and a Russian would say to me, "Aren't you just saying that we should be more like the United States?" my reply was—and not just once, but many times—"Don't confuse the way that most of the world is going with the United States specifically. The ideas of rule of law, of democracy, of openness, of fairness in the way scientific funds are distributed and things of that sort are not uniquely American. In some cases, you could say that America got there first, in other cases you would say that America didn't get there first. If those ideas belong to anybody, it's what you might call the Western community, Europe as a whole. So those are not just American ideas. And you shouldn't think that we're just trying to press America on you. If you ask me to criticize the United States in some instances where it's failed to live up to those ideas, I can do it in a flash! So the ideas are bigger than the United States. Much bigger."[38]

A harsher point of view was presented by Stephen F. Cohen in his book *Failed Crusade: America and the Tragedy of Post-Communist Russia*.[39] Cohen argues, correctly I believe, that the US programs of "democracy-building" in Russia of the 1990s—training in constitutional law, political parties, bicameral legislatures, all the physical trappings of democracy, not to mention economic "shock treatment" that was supposed to introduce liberal, Friedmanesque values and practices—were fundamentally misguided. To put a point on it, in my own words, the notion that we could remake Russia in the American image was arrogant and fatuous nonsense, and it came back to haunt us and bite us in the derrière when the infatuation had passed and the smoke had cleared from the economic turmoil of that time. Cohen published that book in 2000, and subsequent years have shown just how disastrous these efforts were. However, to acknowledge this is not to discredit or devalue the basic motivation of promoting democracy and open society worldwide; it is just to comment on how one goes about it, to caution about what happens when you try to do it overnight.

The Bottom Line

For the entrepreneurs and corporate participants, many of the motivations discussed above may have been on their minds, but the main motivation was clear, without question: making money. Whether the company was a megacorporation or a fledgling small business, the question was always, "Is this going to help advance my company?"

For Randy Guschl of DuPont and Dave Bell of Phygen, while the initial contacts they made in the FSU science community were interesting and helpful, often it was the element of surprise and discovery that made the difference in unanticipated ways. Guschl tells of how DuPont's first foray into FSU science in their traditional core business, chemistry and materials, was not all that impressive, and what turned out to be really useful was in a newer but growing area for DuPont: agriculture.

> We started heavily in material science and catalysis, and we had some interesting successes. They weren't huge home runs, . . . but we kept reminding people at DuPont that there's also a human side here, that for $5,000 they're getting full-time PhDs with staff working for you, so in some cases we were leveraging that impact to do some things that were of interest of DuPont. . . . But there was still reluctance [on DuPont's part].
>
> As we were doing this, the Life Science Department of DuPont, which was growing at that time, also showed an interest. It was a back door, because we, like other companies, were mainly into life sciences through plants and crops, looking for new insecticides, herbicides, and fungicides. It used to be about getting small amounts and spraying it on plants in a greenhouse to see if it had an impact. You literally had to spray tens of thousands of compounds to find something that had activity so that you'd isolate the active ingredient and try to change it.[40]

Through their participation as industry partners in the Global Initiatives for Proliferation Prevention (GIPP) program, however, the DuPont specialists found something that changed their approach: enormous libraries of chemicals, fertilizers, and seeds, that the Russians and Ukrainians had diligently collected (reminiscent of the incredible botanical collections mentioned by Peter Raven and the botanical richness described by Maia Alkakhatsi). Instead of (or in addition to) the inductive approach of spraying plants to see what happens, these libraries made it possible to study the biology and chemistry first and then make decisions about what product lines to pursue. And with that introduction, courtesy of a US government–funded nonproliferation program, DuPont decided to continue the collaboration on chemical and biological libraries on its own nickel, as part of its corporate strategy. Guschl continued:

> So between us and the [former] Soviet Union, we were creating a dance where the Russians were gathering their samples and creating libraries, and we were developing programs to look at their libraries, getting access to five and ten milligrams at a time so you could look at it. . . . And as you would see

a particular library and a particular institute that was feeding certain compounds that were more interesting than others, our people even wanted one of these [GIPP] grants, but then they finally said, "We'll just do it ourselves. We'll fund it ourselves. We're going to go over there . . ." At that point, I no longer had any impact because we had created the bridge; it was what we were told we were supposed to do. We were not there to do this forever.

Indeed, DuPont's approach to these cooperative efforts, like that of most other private companies, was straightforwardly and exclusively strategic. If there was a possible business benefit, they were ready to give it a shot, especially since their investment—the travel costs and modest staff time and effort—was so minimal. In return, in this case as in others, they stood to gain much, in terms of exposure to previously unknown technical information, capabilities, talented people, and even intellectual property. The cost-benefit calculation was impressive. "These libraries that we discovered in Russia," Guschl continued,

> were amazing. Not only that there were one or two but twenty or thirty or so, and many of them were public. Then there were ones that popped up that even our government didn't seem to know about. There had been a proliferation of those laboratories which sort of defied finding them unless we had a tour guide . . . [from] the scientific network itself. . . . Our people would say, "I'd like to look at this institute down the road," and the Russian scientists would say, "Well, we'll get you in there." And as we were doing that . . . DuPont discovered other libraries and some of them, which were directed at toxics, biotoxics, had some very interesting genetic material which I don't know for a fact, but it was being tested and looked very interesting at that time.[41]

Dave Bell, whose initial exposure to the Soviet Union derived from his company's purchase of a license for *bulat* metal processing technology from Russia, adapted it successfully for his company. As we saw earlier, at about the same time as he formed his own startup company, Phygen, through a mutual friend he connected with an Israeli scientist who introduced him to a prominent Russian materials scientist from Tomsk (Gennadiy Mesyats), who encouraged him to visit, and as a result he obtained one of the first grants under the CRDF's Next Steps to the Market Program. While Bell's commercial success was most related to his commercialization of the bulat technology his former company had acquired earlier, from the CRDF grant he moved on to work with the US Industry Coalition in the Department of Energy's GIPP nonproliferation program, becoming a member of its board of directors.

One quick observation from these two stories, which are not necessarily broadly representative but which do have this in common: In many cases—possibly, in the most successful and sustainable ones—the commercial interests of for-profit companies in government-funded or managed nonproliferation programs were ultimately the drivers of sustainable and sustained results. In retrospect, it is questionable whether the government's nonproliferation goals—redirecting (or "engaging") former Soviet WMD scientists in civilian research and development—were very durable at all, because after the programs' demise, it is reasonable to believe that most of those scientists returned to the bench at their weapons labs.

Even during the programs' lifetime, the record presented by the National Nuclear Security Administration (NNSA) to Congress about job creation was very modest; we at US Industry Coalition (USIC) knew that, because we did the bean-counting under NNSA's watchful eye. Sustainable or sustained jobs in the FSU, much less in the United States, created as a direct result of the GIPP program were always disappointingly small. The number always hovered around twenty-eight hundred,[42] and as I recall, the number of sustainable civilian jobs claimed as having been created in the United States never exceeded about a tenth of that number. Yet if you ask people like Randy Guschl, Dave Bell, Terry Lowe, and many other former GIPP industry participants how many US jobs they created through their successful commercial ventures, you'd hear about hundreds and thousands. These jobs and truly significant commercial results—enabled, perhaps, by the government program but driven by the profit motive of the US industry partners—were generally not part of the official record presented to Congress, because the causal chain was too indirect and the successes were really those of the companies' proprietary business practices, not the government's actions. And that's the pity of it, for in the case of GIPP, Congress never really understood the program's full impact. But from our standpoint in this discussion, the difference between the twenty-eight hundred jobs reported during the program's lifetime, and the thousands of jobs reported informally by the companies themselves was the difference between "doing OK" and "doing well."

The People

Finally, there is the curious case of Roald Hoffmann. In chapter five I related the counterintuitive story about how this future Nobel Laureate chemist decided to spend a year in Russia as a graduate student in the early 1960s—precisely to avoid making a decision about whether to continue in chemistry or in science altogether. Later, he would visit the Soviet Union or Russia or Ukraine regularly, about every two years. I asked him, "Why did you keep going back? Was it the science that was in some way important for you?" No, he said emphatically. "It was based on personal factors—liking people and their work." And he went deeper:

> [There was] also something more, an underlying feeling that the search for knowledge and for understanding is there everywhere around the world. In the former Soviet Union and Cuba for sure, very different cultures and different situations. In the Soviet case, the Russian case, there is a history of cultural achievement and scientific achievement and talent. The case of Cuba is something else. There is immense native entrepreneurial spirit—which we can see in the Cuban immigrants in the United States. But the island was left with nothing after the revolution; all the middle class had left, essentially. And they had to train engineers and scientists from nothing. This is admirable.
>
> In the Soviet Union, something else appealed to me about scientists; that is, the typical Soviet scientist was a much more interesting person than the average American scientist. And different in ways that matched my interest. Suppose you were an intelligent nineteen-year-old at the end of *gymnasium* there, and you were faced with the choice of going into history, philosophy, or science. In the Soviet Union, every kid would sense that if they went into history or philosophy, they'd have to tell the party lie. You could make your way . . . in some aspects of humanities—perhaps if you chose some obscure field like Asian art, you might be OK—but if you chose anything that had to do with society, you were under party control. But in science things were different, there was this window into the world. . . . Journals were available from abroad. And so what you found in the Soviet Union was kids becoming scientists, who otherwise would have become literature professors. And they kept their passions. There's also a culture, a tradition, of the intelligentsia. I found the Soviet scientists to be interesting people.[43]

So for Hoffmann, the reason he continued to visit Russia over several decades wasn't at all the science, foreign policy, concerns about global security, or the aspirations for world peace or promoting democracy. It was his personal interest in the people. As we shall discuss in chapter 10, the people Hoffmann found so interesting were indeed members of a very Russian social grouping—the intelligentsia—that is essential to a full understanding of the relationship between science and society in Russia, at least prior to 1991. And that is why Hoffmann's comment was not a mere anomaly in my interviews, but perhaps one of the most profound observations of all.

Notes

1. Interview with Kip Thorne, October 6, 2015.
2. Interview with Marjorie Senechal, May 10, 2016.
3. Interview with Paul Hearn, August 5, 2015. The state of Soviet geology was indeed mixed. As Graham wrote in 1993,

Geology in the Soviet Union in recent decades was strong on theory and on traditional methods of observation and analysis, resulting in a large quantity of data, but weak in reliable instrumentation, innovative analysis, quality of data, and computer applications. Soviet geology was characterized by a certain conservatism in interpretation. For example, Soviet geologists were very tardy in accepting the revolution in geology caused by the development of plate tectonics. The main obstacle here does not seem to have been ideological or political in any straightforward sense, but, instead, the authority of a few administratively powerful Soviet geologists who had staked their reputations on opposition to plate tectonics.

On this, see Graham 1993, 233.
4. Interview with Thomas Pickering, September 24, 2015.
5. Carey 1982, 139.
6. Ibid., 140.
7. See also the story of how the United States obtained missile silo technology through the US-USSR Housing Agreement (see chapter 7, "John Zimmerman: Missile Silo Technology"). The early 1980s were a time when Defense Department concerns about Soviet theft of US military technology through the cooperative science programs reached a peak. The crisis led to a 1982 report by the National Academy of Sciences, *Scientific Communication and National Security*, which called for a broad reexamination of US policy on the issue. In a brief detail from NSF at the White House Science Office, I drafted a National Security Study Directive (NSSD), which was signed by President Reagan, calling for such a process. That NSSD led eventually to President Reagan's National Security Decision Directive 189, "National Policy on the Transfer of Scientific, Technical and Engineering Information" (September 21, 1985), which stated that "it is the policy of this Administration that, to the maximum extent possible, the products of fundamental research remain unrestricted."
8. Interview with Marina Tediashvili, December 8, 2015.
9. Also see the section in this chapter titled, "Global Security."
10. Tediashvili interview.
11. Kutter, "About Us."
12. Interview with Maia Alkakhatsi, December 7, 2015.
13. Ibid.
14. Alexander Ruzmaikin, in interview with Alexander Ruzmaikin and Joan Feynman, October 6, 2015.
15. See chapter 2.
16. Pickering interview.
17. Ibid.
18. Ibid.
19. Ibid.
20. Ibid. Indeed, some very important practical results in aircraft design for Boeing emerged from that effort, as we shall see when we look at the commercial impacts of science and technology cooperation in chapter 8.
21. Interview with Norman Neureiter, January 12, 2016.
22. Ibid.
23. Ibid.
24. Memorandum from Glenn Schweitzer, May 28, 2016.
25. Ibid.
26. Interview with Rose Gottemoeller, April 21, 2016.

27. Interview with Laura Holgate, October 13, 2015.
28. Ibid.
29. Interview with Siegfried Hecker, January 28, 2016.
30. Hoffman 2009.
31. Hecker interview.
32. Tediashvili interview.
33. Ibid.
34. Interview with Andrew Weber, April 25, 2016.
35. Interview with Lawrence Crum, January 6, 2016.
36. Interview with Loren Graham, October 19, 2015.
37. Interview with Peter Raven, February 23, 2016.
38. Graham interview.
39. Cohen 2000.
40. Interview with Randolph Guschl, March 21, 2016.
41. Ibid.
42. "Thanks for Responding to 2010 Commercialization Survey."
43. Interview with Roald Hoffmann, February 18, 2016.

7

SCIENTIFIC ACCOMPLISHMENTS

WE NOW TURN OUR ATTENTION TO THE SECOND question of my interviews: How did it go? In this chapter, we will hear from several respondents about what they viewed as the scientific accomplishments of their collaborations. It is important to keep in mind, of course, that the following is a subset of interviews with a very small fraction of scientists who engaged in US-FSU collaboration over a period of many years. That said, they come from a cross-section of several scientific fields and are helpful in illustrating the different ways in which the collaborations promoted the advancement of knowledge.

Kip Thorne: Gravitational Wave Detection

In chapter 5, we saw how in the late 1960s Kip Thorne of Caltech became strongly attracted to the work of Russian theoretical physicists on astrophysics and gravitation and maintained an active collaboration with them for decades thereafter. Early in that period, he also met a Russian gravitational experimentalist, Vladimir Braginsky, from Moscow State University: "Zel'dovich introduced me in '68 to Vladimir Braginsky. In contrast to the others, he was an experimenter. He was a superb experimenter. Along with Joe Weber, he was one of the two founders of the field of gravitational wave detection. He was just tremendously impressive, working with the technology they had there."[1]

At this point it is worth interjecting that to find an outstanding experimentalist in the Soviet Union was, if not unusual, at least counterintuitive. The general wisdom in the West, at least, was that Soviet science excelled in certain areas that were heavy in theory, but not so much in experiment, largely because of the relatively primitive nature of equipment and instrumentation in Soviet scientific laboratories.[2] But this was not the case with Braginsky.

He was completely competitive with people in the West, despite his poorer technology. I became very close friends with him and he became my leading collaborator because he was an experimenter with whom I could talk about experiments to test relativity as well as gravitational wave experiments, an experimenter that I could learn a lot from. I'm a theorist; I've done some experimental design but have never done experiment.... So anyway, by the mid-'70s, Braginsky had become the person who sent most of the invitations to me. I was collaborating sort of half with him and half with other people in Moscow.... It was largely in discussions with him, and with Rai Weiss at MIT, that convinced me that gravitational wave detection was going to succeed.[3]

Indeed, based on this sense of confidence, Thorne, together with several American colleagues, became a principal initiator of and agitator for the Laser Interferometer Gravitational-Wave Observatory, or LIGO. In the 1980s, several attempts to get the National Science Foundation (NSF) to fund this highly speculative and incredibly expensive project failed, until in 1988 the LIGO group received its first award. The project was a huge risk. "It never should have been built," Richard Isaacson, the NSF Gravitational Physics program manager at the time, said in 2016 to a writer from *The New Yorker*. "It was a couple of maniacs running around, with no signal ever having been discovered, talking about pushing vacuum technology *and* laser technology *and* materials technology *and* seismic isolation *and* feedback systems orders of magnitude beyond the current state of the art, using materials that hadn't been invented yet."[4]

In our conversation of October 6, 2015, Thorne continued: "And now, it almost certainly will succeed sometime in the next two to three years. We're now carrying out the first gravitational wave searches with advanced detectors that have sufficient sensitivity to make us pretty confident there are waves."[5]

Somewhat facetiously, I asked him, "So you don't have to wait for two black holes to collide?"

"Well," he answered, "the first detection will be probably two black holes colliding at distances of several hundred million light-years. But Braginsky convinced me in 1975 that this will probably succeed—he and others, principally Weiss, but largely he—and that led me to build to build a research group in gravitational wave detection at Caltech and then to cofound the LIGO project with Rai Weiss at MIT and Ron Drever at Caltech. And then Braginsky became a principal advisor to this project."[6]

In fact, by the time of the interview, the collision of two black holes had already happened three weeks before, on September 14, 2015. It was not publicly announced, however, until February 2016 to allow for careful verification. Obviously, in October 2015, it was not possible for Thorne to reveal to me that what he predicted would happen "sometime in the next two to three years"—probably the most important discovery in physics in one hundred years—had already taken place.

"Getting Braginsky out" to make foreign visits, including spending time with Thorne in California, he continued,

> was not entirely easy. He was not Jewish.... He had the advantages of being a party member, though he did not toe the party line. For example, we got him out to the General Relativity and Gravitation Conference in Copenhagen in 1971, and there, this [international organizational] structure for this field was created, in the form of the International Society for General Relativity and Gravitation. We fashioned it to facilitate Eastern Bloc memberships; you could have member states that joined, or organizations could join, or you could have individuals that join.
>
> There was an international committee that ran these things. Braginsky was elected to it right in the beginning, as was I. At this meeting in Copenhagen, where the society and the committee came into being, there was a big brouhaha over the Soviet Union not having given visas to some Israeli scientists in time for them to go to the meeting in Tbilisi three years earlier. And when impassioned speeches about this were made, most memorably by Ivor Robinson of the University of Texas, the whole Soviet delegation and most of the Eastern Bloc delegation just got up and walked out. This was in a huge auditorium, there were perhaps five hundred people there, so the Eastern Bloc got up and walked out. I went out a few minutes later and talked to Braginsky—we were close friends—and talked him into coming back in, giving a conciliatory speech....
>
> He paid a fairly high price [for that]. He was denounced by at least two famous Russian participants in the meeting after they returned to Moscow and wound up on a travel blacklist for a while. That characterizes Braginsky, that story, that he really had the guts to do that and was willing to do that.... He had the standing to be able to survive it, but he had to give up his ability to travel for a while. This was in 1971. He did come out pretty successfully after that.... He would occasionally be denied an exit visa.... He had a close friend who was on the staff of the Politburo who could look at his record. So when he would be denied, he would just speak to his friend who would look at his file and would say, "It's just the authorities letting you know who's boss. Don't worry; you'll get out next time."[7]

In 1991, when science in the FSU went into a tailspin, Braginsky's laboratory at Moscow State University was hit hard. Thorne was worried that it would fall apart, Braginsky's group would dissolve, and the historic attempt to detect gravitational waves be placed in jeopardy. Then, in late 1991, Rich Isaacson knocked on my door in the NSF's Division of International

Programs. He said that one of his principal investigators had a Russian colleague in a really important project who was about to fall through the cracks. I asked who the PI was, and he said, Kip Thorne. I became very interested because I had actually met Thorne some fifteen years earlier when I was managing the Working Group on Physics under the late US-USSR science and technology agreement. By 1991, a new US-USSR basic sciences agreement was in place, and I had the stupendous (stupendously small) budget of about $135,000 to fund projects. (The NSF budget for the S&T agreement, in contrast, had been in excess of $3 million.) I explained this to Isaacson, who said that even if I could make a small supplementary award of up to $5,000 to Thorne's grant from the Physics Division, that would help a lot. So I did.

In November 2014, after seeing the movie *Interstellar*, for which Thorne had been the scientific advisor, I decided on a lark to write to him to tell him how much I enjoyed the movie. "I'm sure you don't remember me," I said in my email. Within twenty minutes, I got a response from him, saying, "I remember you very well. Among other things, you played a crucial role in helping keep Vladimir Braginsky's superb research group alive at Moscow State University in the early to mid-1990s."[8] I was pretty floored by his response, and it was then and there that I probably got the idea to reconnect with Thorne and others to do interviews for this book. When I met him at his Pasadena home in October 2015 for an interview—my first one—I asked him, "So what happened with that?" Thorne told me:

> When the Soviet Union crumbled, Braginsky had a group at Moscow State University. There was nothing else like it in the world in terms of particular areas of technology development that they were doing for LIGO. He was developing suspension systems to suspend—by fused silica fibers—to suspend these mirrors that swing in response to gravitational waves. It was crucial that the suspension system was as low-friction as possible. And he developed suspension techniques that were lower in terms of friction by several orders of magnitude than any others in the world; he worked that. He was also studying in depth the noise due to sudden jerks, which could be devastating noise for us. So he was really crucial to the future of LIGO and was also a close personal friend. The science groups in the Soviet Union were falling apart fast and the issue for us was to hold this group together.
>
> So the first step, when things crumbled and his group was about to fall apart, was that we got Caltech to provide internal Caltech funds to hold the group together. The second step was you, and then I think the third step after that was Soros, which you were also involved in. And then after that, by then, we had figured out how to launder funds—you could call it laundering, but in reality we developed a mechanism to have an NSF research grant to Caltech with a subcontract to Moscow State University. . . . Soros was crucial in doing

that transfer [through the International Science Foundation's [ISF] innovative Grant Assistance Program], because [at that time] the problem was, how do you do the transfer of university funds?... It worked great and it was crucial to Advanced LIGO which is to say, I don't think it would have succeeded without this.[9]

I was momentarily speechless. "My $5,000 grant led to the discovery of gravitational waves? Really?" I asked Thorne. "And how did they use that money? Did they use it for personal financial support or something else?" He told me, "It was used primarily just for salaries for members of the group, so it could be held together.... That went a very long way in that era. They were paid $300 a month. Or $200 a month. I think Braginsky was getting $300 a month.... So they were able to live on that. Beyond that, they were able to work full time on the project. They weren't going out moonlighting. And so that genuinely held the group together in a very strong form that exists to this day."[10]

I do not think it would be a serious exaggeration to say that this was the most consequential $5,000 research grant award made in the history of NSF's international programs, if not ever.

Julie Brigham-Grette: Understanding Climate Change

Julie Brigham-Grette's 2009 coring expedition to the impossibly remote Lake El'gygytgyn was a groundbreaker in more than the literal sense of the term. It resulted in hard evidence about the impact of ancient global warming due to variations in the Earth's orbit that also provides a clear preview of the impact of global warming due to another factor: anthropogenic activity. The following long interview extract speaks eloquently for itself:

> The results were amazing. Somewhere in the third week of April, we actually hit bedrock. So we had achieved the entire 3.6 billion years of record, then we went another two hundred meters into the impact bed.... This was a ... volcanic rock target that the impact had hit that ... nobody had seen ... before....
>
> I could summarize the results in a couple of major points. First of all, we found that ... the Pliocene, 3.6 million to almost 2.2 million years ago, was pretty warm, much warmer than today, [which is] no surprise because CO^2 was like today.... There were hemlock and walnut and all these exotic trees up there, evidence of "large Pliocene-Pleistocene megafauna." ... It is very clear that the warmth in the Arctic in those days was driven by very high carbon dioxide.... It's kind of what we're doing now to the atmosphere with human emissions. We can see from this that at about four hundred ppm [parts per million of carbon dioxide], the entire Arctic was forested, there was no Greenland ice sheet, and the other thing that we know now ... is that through

part of that period, there was no West Antarctic ice sheet. Global sea level was high, the entire Earth was warm, and we can document that very clearly. We see the transition into the first major glaciation, so we have the entire story at Lake El'gygytgyn about how the Arctic evolved from a forested Arctic into the tundra that we see today.

The next major point ... which is still quite amazing to me ... [is] that we can document seventeen periods in the last 2.8 million years when the climate was extremely warm—they call them super-interglacials—and as you get into the younger part of the record, these warm intervals are not driven by carbon dioxide. At least for the last 800,000 years, the ice cores in Antarctica show us that carbon dioxide records weren't necessarily high. So we have evidence for super-interglacials that require additional explanation. We have two papers in *Science*, one in 2012 one and 2013,[11] documenting this warmth and the seventeen super-interglacials. . . .

Several of the super-interglacials are now recognized worldwide as really interesting. They'd never been seen in the Arctic before. So to look at what we knew of the geologic record, [for some] 200,000 or 300,000 years, we really didn't have any evidence of the climate evolution in a continuous manner. There are little pieces of geologic time recorded here and there . . . but nothing continuous, not like an ocean record.

And then the other thing that happened in 2009 is that while we were drilling, the Antarctic Drilling Program (ANDRILL) had just published their results from cores recovered from under the Ross Ice Shelf. They were able to document when the West Antarctic ice sheet came and went . . . over time. That work was quite remarkable to the science community, because suddenly it was clear that the West Antarctic Ice Sheet [WAIS] is not stable. The science community had thought that the WAIS had been stable for seven million years and it had never been smaller. Well, guess what? That assumption was not true. ANDRILL gave a different story.

So fast-forward to where we are today: What we think we have, through work done together with our graduate students, is a Milankovich Earth Orbital explanation for why these super-interglacials in the Pleistocene happen when they do when CO^2 is not as high as today. It's quite complicated but it turns out that changes in the shapes of the Earth's orbit around the sun and the tilt of the earth and then the season in which we have perihelion and are closest to the sun—so this is procession and tilt and eccentricity, all of those things—conspire, due to gravitational influence, to alter the duration and intensity of summer in the southern hemisphere and in the northern hemisphere. That is, that when eccentricity is really, really round but the tilt is high, it causes the two hemispheres to link up so that instead of being out of sync because of the tilt, they actually line up. . . . So the poles can be out of sync and then suddenly, when you get this configuration, it causes the two summers to link up. We think every single one of these super-interglacials happens when you get the synchronization of the two hemispheres, and it causes the Antarctic to stay in an interglacial mode. A prolonged interglacial in the southern hemisphere causes the demise of the West Antarctic ice sheet and then, the next time you get an orbital high in the northern hemisphere, you get extremely

> warm super-interglacials because of the alignment. It's just a theory we are still trying to work out....
>
> If you ... look at what it means, this record in Earth history tells us that Greenland is very vulnerable to small changes in climate, causing the ice sheet to come and go, and likewise (based on ANDRILL), the West Antarctic Ice Sheet can come and go, much more easily and more often than we ever thought.... And so now, here we are today with raising global temperatures forced by an atmosphere with CO^2 above four hundred parts per million; we're heading—really heading—back to Pliocene time three million years ago, and this is not a super interglacial. Rather, the forcing of the climate today is driven [by] CO^2. So it's no surprise that the science community is saying that we're experiencing an unstoppable retreat in the West Antarctic Ice Sheet right now. The geologic record and new observations over the past year that came out tell us that, and, of course, now we're seeing Greenland starting to fall apart.[12]

While the expedition's results were impressive, even historic, the project's promise was not evident to some at the outset: "You know I can remember, one [NSF] program manager ... stomping out of the room saying this was a stupid idea, when I first [submitted my grant proposal]. He said, 'Why would you study a lake in Northeast Russia so far away from the North Atlantic, it's not going to have any story?' And yet, when we did those initial cores we could see, 'Wow, it matches the Antarctic ice cores, we can see stuff from Lake Baikal that matched with our lake, it was not a fluke.' It had a global signal."[13]

Tengiz Tsertsvadze: Combating AIDS and Hepatitis C

Tengiz Tsertsvadze is the national AIDS coordinator for the Republic of Georgia. When I met with him in Tbilisi in December 2015, I asked him what kind of success he experienced in his collaborative research with the United States, which took place largely through Civilian Research and Development Foundation (CRDF).

> Big success! ... The Georgian AIDS Center is designated as one of the best in Eastern Europe. Georgia is the first country in the former Soviet Union that provided universal access to antiretroviral treatment for all HIV/AIDS patients since 2004. All HIV-infected patients that are diagnosed are offered treatment free of charge and according to international standards. It is still possible only in Georgia.... As a result of this ... and other achievements, the WHO gave us the highest award in 2009, the Dr. Lee Jong-Wook Award, named after Dr. Lee who was the former director general of [the] WHO....
>
> I believe that if we did not have the opportunity to take part in joint projects, Georgia would not have been able to implement services at this level first for AIDS patients, and later for viral hepatitis cases.... The first five

grants provided by CRDF were extremely important. We received other grants later and some were greater, but the initial five were significant. . . . Later, we received two BTEP[14] grants. The amounts were greater, but the significance of CRDF grants was high since they were first and it was the beginning of cooperation. . . . Later there were other grants. We currently have four grants from NIH received last year and this year.[15]

For a grant manager like me, a track record of subsequent competitive grants is one of the best metrics of the quality and achievements of a research grant. The Georgian group's subsequent work has been very impressive:

> Last year, in 2014 an international workshop was organized here for participants from the United States and countries of the former Soviet Union. A large delegation from NIH came. . . . There was a major workshop last year and we celebrated the thirtieth anniversary of our center, the AIDS Center. All these achievements, our ability to create a modern AIDS Control Service, which was named the best in the FSU, are the direct result of our cooperation with American scientists. . . .
>
> We constantly had joint projects. More than thirty young scientists participated in long-term trainings in the US after me. These trainings that lasted between one and three years. The areas they studied included public health, epidemiology, clinical disciplines, and lab work. There were more than thirty scientists alone from my center, but if we were to include other institutions, there were probably twice as many. Unfortunately, half of them stayed to work in the US, but the second half came back and works here. They created this service. These connections helped us a lot, and the foundation for all this work was laid by CRDF projects. . . .
>
> Currently, as a continuation of these achievements, Georgia was chosen to be the first country in the world to work on a historic project aimed at elimination of hepatitis C.[16]

Marjorie Senechal: The Topology of Crystals

Mathematician and geometer Marjorie Senechal of Smith College did her collaborative work in the Soviet Union, and later Russia, not at one of its famous mathematical institutes or university faculties, but at the Academy of Sciences's Institute of Crystallography. Her first host was Nikolai Sheftal. As she told me in her interview, "Professor Sheftal . . . was sweet and kind and wonderful as he could be, [but] we never really were able to understand what each other was saying. It wasn't a language problem; it was that I'm a mathematician, [and] he was really a crystallographer. He was a very good crystallographer, and his field was crystal growth, which is what I was interested in, but he didn't see it the way I did, even though his paper was the one

I got so excited about. We didn't quite mesh there. But in his lab, there was Eugene Givargizov."[17]

At the mention of Givargizov's name, I must have jumped, because I remembered him so clearly from an exchange visit he had made some forty years earlier when I was working at the National Academy of Sciences. Yevgeniy Givargizov was unlike almost any other Soviet scientist I had met at that time, first of all because of his exotic ethnic heritage: He is Assyrian, which was a great oddity for me, and he gave me a beautiful small sculpture of Assyrian design (of a value of less than twenty-five dollars, so I did not have to report it) that I thought belonged in a museum and not on my desk. I also remember him as a very courteous and delightful person. She continued:

> Givargizov was in that lab; he was actually head of the lab, because Sheftal had officially retired, though he still came to work every day. Givargizov knew lots of people in the mathematical crystallography area, and he thought that there was one I should meet, and so he brought him up to meet me, a youngish—not much younger than me, but at that time I was younger—crystallographer named Ravil Galiulin, who was Tatar.
>
> Galiulin was ferociously sensitive about being a Tatar. I remember sitting with him in a lecture, much later; someone was talking about Pasteur's work on crystals of tartaric acid, and all Ravil heard was "Tatar." "WHAT ARE YOU SAYING ABOUT TATARS?" he yelled. I said, "Be quiet, Ravil!" and he calmed down. But anyway, he brought me all the papers by himself and several other mathematicians.
>
> He was at the Steklov Institute and at the Crystallography Institute—he was at both—and worked with Boris Nikolayevich Delone. Delone was a mathematician who, in the '30s, had conceived of a whole new way of understanding crystal structure: rather than by looking at symmetries, looking at local formations and seeing them grow. I loved this work. It changed my whole view on what I was doing, how I was doing it. Delone's family had been originally been French . . . an ancestor came with Napoleon and stayed. Delone's work is magnificent, and I'm still working on that and writing papers based on his ideas. Both Delone and Galiulin have died, but some of [Galiulin's] other students are still alive and active and we're in touch all the time. . . . It's a big part of my life, professionally and personally, too. It's been absolutely wonderful.
>
> So that is how . . . everything fell into place. Galiulin and I wrote one paper together that we translated, a major paper by Federov, and we published that later. But more importantly than any of that is that just my whole way of looking at things was changed completely. And when there was a big shake-up in the field of crystallography in the 1980s, when so-called quasi-crystals were discovered, I was in [a] position to realize Delone's approach to crystal structure was the tool that we need now to use to understand these strange new things, not to be looking in the more formal group theoretical stuff in the way

that crystallography was seen before. And that has become the way of doing it. I'm oversimplifying, but that's basically it. And so I feel I played some role in getting that word out and getting that material used, because people outside of Russia didn't know about that at the time.[18]

Siegfried Hecker: The Excitement of Creating New Scientific Knowledge

In chapter 4, we read about the historic post-1991 efforts of Los Alamos director Siegfried Hecker and others to stem the spread of technologies and materials related to nuclear weapons through joint work with the closed Russian nuclear laboratories. There was a great deal of basic scientific collaboration that was an essential part of this relationship. Much of it, however, was in rather unconventional fields that were not typical of, or accessible to, ordinary physicists, as he described in his interview:

> In 1992 the Russian and Soviet lab scientists talked with us about the science. There was a guy by the name of [E. E.] Meshkov, one of the great Russian scientists—we had known his name, but we had never met him—of Richtmyer-Meshkov instabilities....
> So we listen to those guys... [who] had people and capabilities that were incredible. In our business, they were as good as anything; we had in some areas better [equipment]. We had better computers, we had better electronics. They did parallel processing already by that time. They picked up these crappy computers and had parallel processing algorithms before we did. They had mathematics, and they didn't have the big computers. They told us. "You guys are lazy; you just use the number-crunching computers.... We have to think!"
> So I said, "Hey, we can work together. We have fast machines, you have better algorithms; we can work together." In this magnetic accumulation stuff, they had better explosive systems to be able to get to high pressures; we had better diagnostics. So we laid it out at that time, and we said, "You know what we can do together? We can create new science.... We can do the science that's the essence of the fundamental science that helped you understand nuclear weapons. And if it's fundamental, we can do it together."[19]

In his two-volume work documenting this relationship, Hecker writes the following about the purely scientific collaborations that took place under the lab-to-lab program:

> A ... critically important aspect of scientific collaboration was that scientists on both sides longed for doing something new, constructive, and exciting together. Washington's primary driver for engaging the Russian nuclear complex was to prevent bad things from happening.... However, Russian scientists, much like American scientists, want first and foremost to *do* things,

not just prevent things from happening. In other words, the excitement is in creating new scientific knowledge of developing new technologies. The attraction to do together the new and exciting things . . . built the relationships that paved the way for tackling the most sensitive measures that emerged at the Cold War's end.[20]

Even though the research was on the nature of basic physical phenomena, it did raise the eyebrows of senior US government officials, and possibly also the Russians':

But then we also laid out, during that first visit in February 1992, about eight areas: nuclear with safety nuclear materials, protection control and accounting, antiterrorism, antiproliferation, basic science together, even ecological and environmentally related science, we laid them all out. I came back to Washington with John Nuckolls, we presented [it] to Admiral Watkins, and he said, "Well, you guys got out ahead of your headlights. You can do the basic science; these other things, you know, are off limits." When I was told that, when our list that we put together—we signed a protocol—we said, "Hey, look at this; these are the areas of interest." So we went to the NSC [National Security Council, to appeal Watkins's decision] and the list [that they had agreed on with the Russians] was thrown in the wastebasket and said not to exist because we were not to sign anything like that.[21]

That, however, did not deter the US lab scientists:

And then actually over the next ten to twenty years, that's what we did. We worked on those things and then [did] some work on exactly what the government wanted us to do. Some of that we did from lab to lab, some of that the governments had us do, but they involved the lab people too. So then over the next twenty-some years, we involved more than a thousand laboratory people on each side, and we didn't just go just sit at a conference table or sign a piece of paper; we had our people over there behind the fence at these facilities doing the six positive experiments for highly magnetic fields, looking at superconductors in high magnetic fields, and they came to Los Alamos.

So when you ask, "So what did you do in scientific research?" that's what we did. Even if there were nothing else. . . . There is an unusual area called pulsed-power, high-energy density physics—that's what's important for our underpinning science. The Los Alamos and VNIITF[22] guys worked together, and over these twenty-some years, they had four hundred publications and presentations—four hundred joint publications in the best international journals. Absolutely fantastic work. So that was one product. They didn't only create new science; they created new ideas. Then we also worked in various technologies and tried to transition them to civilian research. . . . So we have two chapters on those topics in the book.[23]

Four hundred publications in the best international journals is not a bad indicator of scientific impact.

On the practical side, Hecker mentioned in passing one example of how direct scientific expertise made a critical contribution to decommissioning nuclear weapons in Siberia: The Russians could not figure out how to take them apart as a step to destroying them:

> They asked us, "How do you get these weapons apart?" When they come back from the field, they age, they change. High explosives are terrible stuff. Plutonium's very difficult stuff.
> They said, "We've got all these weapons but we're having trouble getting them apart."
> And a guy from Livermore said, "Oh you know, DMSO, dimethyl sulfoxide. It's a chemical solvent and it'll just dissolve the interface, and then you take them apart."
> Those guys, they were amazed. They said, "You've given us a gift from heaven."[24]

A common chemical solvent—dimethyl sulfoxide, a byproduct of the wood industry, known for decades—did the trick. Why hadn't the Russian scientists thought of that? Why did it take an American scientist from Livermore to figure that out? I don't want to draw too much from this one small episode, but it strikes me that one possible factor was the very nature of the Soviet and Russian science system and its culture. For all the theoretical depth in Soviet physics, both in educational traditions and research, once Soviet scientists got into their institutes and jobs, there was surprisingly little cross-fertilization between scientific disciplines. Even within a given discipline, such as physics, there were a plethora of Academy of Sciences institutes devoted to intensive study of specific subfields: theoretical physics, general physics, nuclear physics, solid-state physics, and so on. True, there were institutes with cross-disciplinary names, such as the famous Institute of Chemical Physics in Chernogolovka, which was particularly well known for its work on explosives.

But by and large, the style of interdisciplinary research, which became increasingly common and indeed the norm in many Western countries and particularly in the United States beginning in about the 1960s, was not at all typical of the Soviet tradition. In the West, interdisciplinary research was born in research and teaching universities, which were not chopped up into tiny subdisciplinary pieces. On the other hand, in the Soviet Union, the highly vertical, segmented structure of science, embodied in huge, specialized institutes, discouraged interaction between disciplines and instead encouraged a type of overspecialization that may have actually hindered scientific progress in some ways—not to mention the disassembly of nuclear warheads.

Terry Lowe: Nanotechnology

Terry Lowe, like his Los Alamos colleague, Sig Hecker, had had some exposure to Soviet research, in his case in materials science (Hecker was also a materials scientist), prior to the USSR's breakup in 1991. And like Hecker, he became deeply engaged in the lab-to-lab program and nuclear nonproliferation efforts.

In the course of those contacts, Lowe met a materials scientist from the Ufa State Aviation Technical University, Ruslan Valiev. Valiev was doing basic research on nanostructured materials, which of course had military applications for airplane fuselage and wing coatings. Lowe was interesting in testing these materials in large-scale simulations that were being done at Los Alamos. When he visited Valiev in Ufa, he was intrigued by what he saw:

> I never found institutes that were as smooth and easy to work with as Ufa State Aviation Technical University. The other institutes were fine, but Valiev and the people in Ufa really knew how to work in the international community, so that was actually a very lucky choice on my part, I suppose, to find someone that was so internationally inclined.... If you try to find him at home, he won't be there because he's always in Japan or Korea, or Germany or China. Or the USA or Brazil. Everywhere. He's always on the road. [But] it's partly because of Russia itself: because the tools that they had for doing science and technology weren't as available in Russia as they are at other major research institutions—for example, the highest-quality high-resolution transmission electron microscopes—those didn't exist in Russia or the Soviet Union in those days. So he had to go elsewhere to do his best work and so he went to other places to access the best scientists and equipment in the world. So I think that's a part of it. And he was an excellent collaborator, but he brought that international character to his institute as well.[25]

This research partnership was the beginning of a long-term collaboration between Lowe and Valiev's group in Ufa. Eventually, Lowe would leave Los Alamos and set up his own small business, Metallicum,[26] which sought to market commercial, civilian applications for the technology. It was in that capacity that I first met Lowe, when he was the US industry partner for a collaborative project in the Global Initiatives for Proliferation Prevention (GIPP) program. The project was probably the most scientifically and commercially productive one in the program during my six years at USIC. Lowe attributes a great deal of credit to the scientific quality of the Ufa group's research and the personal drive of Valiev:

> I think what was really key is that Valiev and his group really knew how to perform. They were great researchers. Valiev may be the most highly published

Russian scientist in all history. With respect to the top ten publications in all of at least material science, six or seven are his, or ours; some of mine might be part of this as well, but we really did something remarkable. And in materials science, if you just look at the top publications, period, I think that [the top] four of all time are [our] technology; all came out of this program. . . . This whole thing just continues to grow. . . . Papers by Valiev rank in the top five of all time for the leading materials journals.

In support of this point, Lowe provided me[27] with several tables, based on bibliometric statistics, showing the ranking of Valiev's publications. While I have not reproduced them here due to space limitations, they are very convincing in support of Valiev's extraordinary publication record.

Lowe continued in his interview:

> The next thing that's really significant is that in 1999, we organized a NATO Advanced Research Workshop, which was held in Golitsyna—Moscow, basically. And that was the biggest turning point for the technology. At that time there were 197 publications on this technology area in the whole world. And after that, it started to skyrocket, and now it's in the tens of thousands. We put together a book from the NATO workshop, which has now been reprinted again, because it was recognized as a big turning point. The workshop included early work that we had done together through the GIPP project, from 1994 through 1999.

The cumulative effect of this collaboration on Lowe's scientific research and on his business, which he started when he left Los Alamos in 2001, was profound:

> So . . . what were my expectations? I just go back to this: I had no idea this would be so successful as it has been. And it has, for me, been sort of the trademark of my career. Prior to that, I was a modeling and simulation person. And after that I became a nanostructured metal person. I have a passion for nanotechnology anyway, which was just growing in the world of science at the same time. The first usage of that word showed up around the early 1990s. It grew in Europe and in the United States, which had the National Nanotechnology Initiative, which started in the Clinton administration, and which is ostensibly the most successful science initiative of all time. . . .
>
> I didn't expect us to make as much progress as we did, and of course, we continued to shift the work directly in commercial directions, not just scientific. But this is what we were supposed to do, this was the promise of GIPP. So we made that promise real. We took our commitment seriously. And everything we did was focused toward commercialization. We actually started filing the first patents disclosures around 2002 to 2004, and we focused on making commercial scale technologies. And today, sitting a mile and a half from where I'm sitting right now, there's about $14 million of machinery that came out of our endeavor. We have pilot-scale manufacturing equipment and we have realized millions of dollars from it. There are over forty to fifty

companies that have looked at the technology, and there are, today, various companies that are looking into adopting that and putting it into products. There are already companies that have it in products.²⁸

John Zimmerman: Missile Silo Technology

John Zimmerman, while serving on the Soviet desk at the State Department and watching over affairs related to scientific cooperation, shared the following story with me:

> One time I got a letter that said, "Just wanted you guys to know I just received a major contract for movable shelters on missile silos in North Dakota where we house the Minuteman Three with those big huge slabs of concrete that when you push the button they slide aside pneumatically unclear and the missile comes out.... I got that largely based on research work at the most basic level that I can document under the Housing Agreement, the US-Soviet housing agreement. I thought you'd just want to know that because of North Dakota's climate, to build and have this structure survive, I got the technology and the know-how from Russian experience of building similar building concrete structures in Siberia. And by working through that cooperative agreement with the Soviet scientists I got this contract."²⁹

I've not been able to track down information detailed enough to verify this claim, but it is very likely that the Housing Agreement had a working group on cold-weather construction. Obviously, it was something with which the Soviets had a great deal of experience. It was another example of how the dreaded loss of sensitive technology to the Soviets decried by Frank Carlucci in 1982 was actually a two-way street.

A Note about Space Science

When one thinks about US-Soviet scientific cooperation, one of the most widespread and enduring images is that of the American *Apollo* spacecraft docking with the Soviet *Soyuz* in orbit over the Earth in July 1975. And going back further to 1957, it was another space orbit feat, *Sputnik*, that shocked the United States into realizing that we were being challenged broadly across the scientific and technological spectrum by a hostile Soviet Union and that we were perhaps losing. But was *Sputnik* a scientific achievement? Most knowledgeable people would say, no, it was an engineering achievement and a remarkable one, most likely an indigenous one. It was testimony to the ingenuity, hardiness, and vitality of Soviet engineering and problem-solving in a very impressive way. It was probably not, however, space science.

In my interviews, I received essentially the same feedback about subsequent joint space missions such as Apollo-Soyuz.[30] It was an engineering accomplishment, with little relationship to basic scientific research. To be sure, one had to plot flight paths with incredible precision to achieve this feat, but this kind of mathematical modeling had always been a Soviet strength. And to be sure, important scientific data came out of this and other joint space missions, especially about human biology in microgravity or zero-gravity conditions, which was critical to future space flight programs.

Working in the National Security Council of the Clinton White House, Rose Gottemoeller devoted special interest and emphasis to scientific cooperation. The International Space Station (ISS) and the *Mir* shuttle craft were some of the issues that attracted her attention in that position. She recalls that she thought that these projects would form the basis of a "thriving scientific cooperation" and that she always wondered why they were both so successful and so "modest" in terms of actual science cooperation. I suggested that these were not really scientific research projects at all but feats of engineering, and she took the point.

Margaret Finarelli, who had worked on international scientific cooperation in the White House Science Office in the 1970s and who in the 1990s and onward served in senior positions at NASA, in which she was directly involved in the development of the ISS, wrote this to me in a private message: "ISS is really different. It was designed as a program to stick it in the Soviets' eye by demonstrating US leadership (both technological and political). By the time we got around to inviting the Russians to join in the early 1990s, there was a whole different thing going on. Can you call ISS successful? Absolutely. And certainly bigger and more visible than any of the cooperative science efforts."[31]

By the 1990s, truly, the entire situation had become very different. The US space program had been deeply injured by the 1986 *Challenger* disaster, and by the time the Soviet Union broke up, the United States found itself without spacecraft and boosters to lift astronauts into outer space to work on the ISS, in which the United States had already made a huge investment.[32] Private-sector space flight was still a distant vision. So US-Soviet space cooperation turned to very practical matters—seats on Russian space modules, boosters, and other commercially based cooperation. A new climate emerged, in which multinational companies like Boeing, for example, could have a new presence in Russia to take advantage of the work of

Russian specialists and to develop new markets for their products. The Boeing 737 became a common addition to many post-Soviet airlines and I was always happy and relieved to fly on one during my many internal flights in and about Russia.

Impact on Scientific Infrastructure

In some cases, the scientific impact of collaborations was felt less in direct research results than in infrastructure. This was especially true in the post-1991 period. In the Soros and CRDF programs, when we made competitive grants to FSU scientists or mixed teams involving FSU scientific partners, fully one-half of the funds going to the FSU side were used for equipment and instrumentation. Salary (or, as we called it, "individual financial support") assistance was certainly important, but we quickly came to understand that the most powerful attractor for scientists to continue to work at the bench was the opportunity to work with modern research equipment. Often, this meant not much more than purchasing the latest models of desktop and laptop computers, but there were also special programs, such as CRDF's Regional Experimental Science Centers program, that focused on providing large or expensive pieces of equipment intended for shared use by multiple institutions in the same geographic area.

Instrumentation and equipment were especially important in the context of science in the FSU. One of the most observations by foreign visitors to Soviet and post-Soviet scientific institutes had always been astonishment at both the primitive state of the equipment and the ingenuity of the scientists in using it and keeping it running at all. Most of the legacy equipment in these laboratories, moreover, was of Soviet manufacture, in many cases, one-off devices fabricated in the institutes themselves. Thus, when in 1991 there was suddenly no money for researchers' salaries, much less even spare parts for exhausted or broken equipment, the internal, infrastructural crisis of science in the region was acute. Foreign programs, including those from the United States, played a key role in reequipping the laboratories, giving them the modest funds needed for spare or replacement parts, and purchasing consumable research supplies such as enzymes and other substances. Without this critical support, it is doubtful that many of the good laboratories that did survive would have continued to function at all.

In many of these laboratories, the visitor can see the evidence of these efforts by looking for the small plaques on many pieces of equipment attesting that they were purchased under one foreign program or another. In the case of US government–funded programs, it was sometimes necessary to send durable equipment to the FSU not as outright gifts but as long-term loans. Other programs simply required that the recipient post a permanent notice on the item saying that it was a donation and often also displaying an inventory number for identification purposes.

These gifts, or donations or loans, were enormously helpful. We have already heard from Georgian botanist Maia Alkakhatsi about how it was through collaborative projects that her laboratory was able to upgrade essential equipment, such as computers and electron microscopes. This story has been replicated hundreds, if not thousands, of times in research labs and institutes throughout the FSU.

A special case, with which I had some involvement, was the roof of the Komarov Botanical Institute of the Russian Academy of Sciences in St. Petersburg. When Peter Raven first visited Komarov Institute in 1975, it was clear that the building was in serious disrepair.[33] By 1991, the roof was about to fall in, with potentially disastrous results for the institute's unique and irreplaceable herbarium, a world treasure. When Raven was contacted in 1992 by Alexander Goldfarb, who was working with George Soros on setting up a multinational US-FSU board of directors for Soros's new International Science Foundation, he was intrigued:

> One day in the early '90s, I received a telephone call from Alex Goldfarb, who told me that billionaire George Soros was organizing a foundation to help with Soviet scientific community survive times that were truly horrible economically. Alex invited me to a meeting about the new foundation, and we were soon underway. . . . I was interested in the cause, which seemed right to me, and also wondered in a vague way if helping the new foundation might in turn help me find the funds to restore the important herbarium building at the Komarov Institute.
>
> Later on, my hunch turned out to be right. . . . When I began to inquire about a grant for the Komarov restoration, I spoke with Valery Soyfer, who was Mr. Soros's key advisor in the matter of charitable giving in the former Soviet Union. Valery consulted the director of Kew and then recommended to Mr. Soros that he authorize a grant for the purpose. We received $500,000 of the estimated $1.15 million cost to fix the roof, walls, and the heating system for the building. I was able to get another $500,000 from USAID through Colorado senator Tim Wirth, who has an interest in matters biological—he is a close friend of Paul Ehrlich's. The rest of the money I raised from friends who wanted to help, even though they had no deep interest in the matter. I was very pleased that the work could proceed, supervised by a Finnish firm, and all the more so that it proved possible to manufacture all the necessary parts in Russia—even though that had by no means been certain when we started the work.[34]

Indeed, in many cases, infrastructural support that came through cooperative projects made the difference between life and death for many institutes—and in both countries, as a matter of fact.

In Kyiv, one of the most innovative research institutes in the field of materials science is the International Center for Electron-Beam Technologies, or ICEBT. ICEBT had its origins in academician Boris Movchan's world-class research in electron-beam vapor deposition, with applications to super-hard materials coatings, electric power transmission, medicine, and agriculture. I had met him in 1980, when he was one of the leaders of the Working Group on Electrometallurgy and Materials under the US-USSR science and technology agreement. After 1991, ICEBT participated in a wide range of projects, including direct support from the ISF and cooperative research projects funded by the US Department of Defense (DOD), the nonproliferation-oriented science centers, GIPP, and CRDF. An early meeting with DOD representatives resulted in a list of research topics of mutual interest to both sides. When I saw Movchan in Kyiv in December 2015, I asked him what he considered the ultimate results of his long history of cooperation with US scientists. "I would put it this way," he said, "very briefly, if we had not had this first little list that was a direct agreement between the US Department of Defense and the ICEBT—if there had not been the support—the international center would not have existed. Why? For the duration of all these years, twenty-one years [as of 2015], we received not a single *hrivna* [the basic unit of Ukrainian currency] from our government. We had no government contracts."[35]

In April 2016, Movchan was honored by the American Vacuum Society's Applied Surface Engineering Division by receiving its R. F. Bunshah Award in recognition of his "pioneering work on electron-beam deposition, and his leadership as an educator and mentor spanning six decades and three continents." This international recognition was testimony both to the quality of his research as well as to the impact of support through collaborative research during difficult times from the United States and the international community.

In his interview, Larry Crum illustrated how the impact of bilateral cooperation with Russia was not a one-way street. He explained: "We thought we were going to have to shut down this unit, which is a part of another unit of the university. But these Russians have been so good and so active and so aggressive that they now essentially have as much funding as any of our other colleagues here. They have saved the department and their reputation has greatly improved that of our own department."[36]

Indeed, while the fear of a massive "brain drain" did not turn out to be of the dire proportions that many feared in the early 1990s, many talented former Soviet scientists, both young and old, did populate many science departments in US universities. (The United States was in third place with regard to the scientific emigration from the FSU; Germany and Israel were the major beneficiaries.[37]) In many of these cases, a common pattern was what Dr. Irina Dezhina, a prominent Russian science and technology policy analyst, has called "pendulum migration": the scientists move back and forth between the United States and Russia, Israel and Russia, and so on, spending a semester to six months in one place and the balance in the other.[38] In other words, these scientists have become part of the world community of scientists, an international fellowship that, according to some, "knows no boundaries." If only that were literally true.

CRDF's Basic Research and Higher Education (BRHE) program was a special case in terms of its impact on scientific infrastructure.[39] Its stated purpose, after all, was to change the structure of Russian science itself, from a bifurcated system in which the Academy of Sciences had primacy in research while universities—with a few exceptions in Moscow and St. Petersburg—were little more than pedagogical institutions. It can reasonably be argued that BRHE had a major impact in that it introduced the concept of the modern research university and how to achieve it to Russian policymakers and universities alike, a path that has since become core government policy. That being said, if one were to revisit the BRHE program's Research and Education Centers today and their host universities, one would find little trace of their existence. Loren Graham, who was deeply involved in the BRHE program's conception and oversight, told me,

> One thing that I've learned—and I should have known it before—is that changing institutions is incredibly difficult. Changing attitudes is easier than changing institutions themselves. Attitudes reside in individual people. I think that BRHE did have a definite impact on individual Russians—the professors, the university administrators, the directors, with whom we directly interacted. We established communication with them and they came to agree, or at least they said they agreed, that what BRHE was trying to do was a good thing and they wanted to work with us to try to do it. But if you go back to those universities now and you look for the traces, for the lasting impacts, it's a little discouraging.[40]

When Graham made these comments in October 2015, there still appeared to be some hope that the heart of the university reforms of the early Putin years—merging two or more universities in selected cities or regions with the intention of producing superior higher education institutions on the general model of the modern research university typical of Western countries—would continue.

However, change was already in the air. Harley Balzer, who with Graham and me and others (including Marjorie Senechal) was among the architects of the BRHE program, had kept in touch with Andrey Fursenko, who, as minister of education and science, had been a key supporter of BRHE in earlier years and built on it to develop an across-the-board policy of strengthening the Russian university system. Writing in a *New York Times* op-ed on July 22, 2015, Balzer reported a recent conversation he had in Moscow with Fursenko, now an intimate advisor of President Putin and a personal target of US government post-Crimea sanctions. Balzer recalled that Fursenko had not only been a strong supporter and enabler of the CRDF BRHE program but subsequently had even asked for CRDF's assistance in developing four additional Research and Education Centers on the same model, to be funded entirely by the Russian government. But by 2015, no doubt embittered by his inclusion on the US sanctions list, Fursenko had changed his tune. Balzer wrote: "Earlier this month, I saw Mr. Fursenko again. I expressed my concerns over the Kremlin's recent actions [in Ukraine]. He told me bluntly that things have changed. He said that this was because 'America cannot tolerate any partner who does not behave like an obedient child listening to a parent's strictures.' Russia, he said, is tired of this."[41] (So much, again, for the realpolitik theory that scientific cooperation can be an "incentive for restraint" by a foreign partner.)

The icing on the cake came the next month. In August 2016, Putin replaced Ministry of Education and Science minister Dmitri Livanov, also a strong BRHE supporter who vigorously pursued the policy of strengthening universities, including merging some of the weaker ones with stronger neighbors into "federal universities," with Ol'ga Vasilyeva, an admirer of Joseph Stalin with close ties to the views of the Russian Orthodox Church on education and science. There was much concern that she would undo the reforms, and in so doing also destroy what was left of Russian science by returning to the Soviet model. And undo the reforms she did.

On September 26, TASS reported that Vasilyeva had decreed that "the merging of universities begun under the previous leadership of Minobrnauka [Ministry of Education and Science] of the Russian Federation will be terminated."[42] So ended Russia's experiment with bringing its higher educational system into the twenty-first century, also an explicit acknowledgment of what had become apparent for years, that Vladimir Putin's regime had abandoned its quest, initiated under former president and current Prime Minister Dmitriy Medvedev, to build on Russia's scientific and technological accomplishments by building a true knowledge economy.[43]

Another area in which some of the cooperative programs after 1991 had a potentially important impact on science infrastructure was in the large-scale implementation and validation of competitive, peer-reviewed research grants. Prior to that time, there was no such thing in the Soviet Union. Scientific research was funded from the top down, from the government, through the academies of sciences, which divided up the funds among its divisions, institutes, and laboratories. While a handful of institutes may have had local competitions for funds (for example, at the Ioffe Physico-Technical Institute, as I was once told by its director, Zhores Al'ferov), nothing of the sort existed on a national scale. The Russian Foundation for Basic Research, established in late 1992 under the direction of senior mathematician Andrey Gonchar, set out to apply such procedures, but its funds were chronically limited by the government for many years to a barely significant level. At about the same time, the appearance of Soros's ISF provided massive support through a competitive grant mechanism and—I would argue, with some bias since I created and managed that system—succeeded in introducing that novel approach (for Russia) on a very broad scale, building confidence and competitive skills among Russian scientists. The ISF was followed in this by a host of other international programs, including CRDF and INTAS. Through these channels, many Russian and other FSU scientists learned how to compete for research grants through a variety of foreign national and international programs. "Today," wrote Graham and Dezhina, "the competitive peer review system is taken for granted by most Russian scientists, and it improves the quality of research."[44]

While bibliometric data on Soviet science are of limited use, especially since during the Soviet period, publication in international scientific journals was relatively rare, some bibliometric analysts in Russia have observed trends that suggest that both institutional reform and international collaboration have had beneficial impacts on Russian science. The data are fragmentary, but intriguing. For example, Valentina Markusova et al. reported in 2013 that the new Russian Federal Universities and National Research Universities have had significantly higher shares of publications tracked by Web of Science,[45] that acknowledge support by an internationally recognized funding agency—50 to 52 percent, respectively, for the federal and national universities versus 42 percent for all higher educational institutions.[46] In the same paper, they observe that "growing international collaboration is a factor of globalization. Russian internationally collaborative papers... received more citations than articled published only by Russians.... Russian research collaboration with foreign countries has a long a turbulent history."[47]

In another study a year later, Markusova et al., comparing the scientific output of the Russian Academy of Sciences (RAS) with that of the federal and national universities, found that "despite the huge financial inflow in the HES [higher education sector] in the last seven years, the RAS performance is still much stronger. RAS is responsible for 56.3% of the total Russian research output and the HES for 42.6%. A significant improvement of RAS in collaboration with HES was observed.... Collaboration with RAS had a significant positive impact on citation score of HES."[48]

In other words, the drawing together of Russian universities and academy institutes, which was a key component of the BRHE program, as well as the government's strong emphasis on strengthening Russian universities, had been good for Russian scientific output. How and whether this trend will continue in the light of the more recent removal of the former RAS institutes to an independent government agency, the Federal Agency for Scientific Organizations, and the halt of efforts to strengthen the university system remains to be seen, but the outlook is not encouraging.

Notes

1. Interview with Kip Thorne, October 6, 2015.
2. This is not to belittle the ingenuity and talents of Soviet experimentalists, whose ability to repair equipment, hand-craft replacement parts, and design entirely new devices of their own manufacture was legendary.
3. Thorne interview.
4. Twilley 2016.
5. Thorne interview.
6. Ibid.
7. Ibid.
8. Kip Thorne, email to the author, dated November 14, 2014.
9. Thorne interview.
10. Ibid.
11. Melles et al. 2012; Brigham-Grette et al. 2013.
12. Interview with Julie Brigham-Grette, December 2, 2015.
13. Brigham-Grette interview.
14. The Biotechnology Engagement Program, administered by the National Institutes of Health and funded by the Department of State.

15. Interview with Tengiz Tsertsvadze, December 7, 2015.
16. Tsertsvadze interview.
17. Interview with Marjorie Senechal, May 10, 2016.
18. Ibid.
19. Interview with Siegfried Hecker, January 28, 2016.
20. Hecker 2016, 2:186.
21. Hecker interview.
22. The All-Russian Scientific Research Institute of Technical Physics, which was known in the Soviet period by its code name, Chelyabinsk-70; it is now also known as "Snezhinsk" for the name of the nearest town.
23. Hecker interview. The book reference is to Hecker, ed., 2016.
24. Hecker interview.
25. Interview with Terry Lowe, March 26, 2016.
26. Now part of Manhattan Scientifics, Inc.
27. Personal message to the author from Terry Lowe, April 30, 2017.
28. Lowe interview.
29. Interview with John Zimmerman, September 30, 2015.
30. Interviews with John Logsdon (November 17, 2015), Rose Gottemoeller (April 21, 2016), and informal discussion by the author with Margaret Finarelli (August 4, 2016).
31. Personal message to the author from Margaret Finarelli, July 1, 2016.
32. O'Rourke (2014) wrote that in 2009 there was a proposal from United Launch Alliance, a joint venture between Lockheed Martin and Boeing, to put "U.S. satellites into orbit aboard all-American Delta IV rockets. ULA presented a paper to the American Institute of Aeronautics and Astronautics detailing how quickly the Delta IV-Heavy could be 'human-rated' (Washington politician-speak for 'safer than sending Christa McAuliffe up in the Space Shuttle *Challenger*'). ULA said 4 1/2 years." However, he continued, the Bush administration got sidetracked with the war in Iraq, to the detriment of investments in the space program: "President Bush said we were going to Mars, and we went to Iraq instead.... And we're a democracy. So we the people share blame for Russia finally winning the space race."
33. See chapter 5.
34. Interview with Peter Raven, February 23, 2016. As the ISF's chief operating officer, I spent two years providing administrative and financial support for the team of Finnish construction experts who were in charge of the roof's repair. It was certainly one of the most tangible projects I had the opportunity to manage in that position.
35. Interview with Boris Movchan, December 17, 2015.
36. Interview with Lawrence Crum, January 6, 2016.
37. Dezhina 2002, 7. The page number reference, here and throughout, is to a freestanding reproduction of the article's text in Dezhina's personal files, beginning with page 1. Neither she nor I have been able to access the article as it originally appeared in *Naukovedenie*, which has no internet presence.
38. Ibid.
39. See chapter 4 subsection titled, "The Civilian Research and Development Foundation."
40. Interview with Loren Graham, October 19, 2015.
41. Balzer 2015.
42. "Glava Minobrnauki pristanovila protsess ob'yedineniya vuzov." My thanks to Harley Balzer for bringing this article to my attention.

43. In a more recent conversation with Balzer, he indicated that Fursenko, who holds the powerful position of presidential advisor on education and science, has managed to undo the worst of Vasilyeva's damage.

44. Graham and Dezhina 2008, 164.

45. Web of Science is an online subscription-based scientific citation indexing service that tracks citations to scientific publications and is a standard reference tool for evaluating their impact.

46. Markusova et al. 2013, 9.

47. Ibid., 7.

48. Markusova et al. 2014, 11.

8

OTHER ACCOMPLISHMENTS

CHAPTER 7 WAS DEVOTED SOLELY TO THE SCIENTIFIC accomplishments described by my interlocutors; however, the advancement of scientific knowledge was but one of the many objectives of the creators and managers of the science cooperation programs described in this book. But there were many other policy goals, such as developing personal contacts between citizens of the two superpowers, promoting US foreign policy and national security, providing emergency assistance to former Soviet scientists after the fall of communism in 1991, and spreading democratic values and institutions. Moreover, as we have seen in chapters 5 and 6, it was a mix of these goals that motivated the participants themselves. Seldom, if ever, was the advancement of science itself the lone objective. Understanding the extent to which these other goals were met over time is thus vitally important to our ability to draw some conclusions, or at least lessons learned, which we will discuss in part 3.

In this chapter, we will consider the views of the participants about the extent to which the programs were successful in addressing these nonscientific, or "policy-driven," goals.

Proliferation of Weapons of Mass Destruction

From Sig Hecker's and Andy Weber's testimony, we have already gotten some insight into the role of scientific cooperation in the efforts of the United States to combat the proliferation of nuclear and biological weapons of mass destruction (WMDs) and their technologies to hostile powers and groups. "That certainly felt good," Weber said of the destruction of "the largest anthrax factory in the world" in Stepnogorsk, Kazakhstan. About the impact of transferring six hundred kilograms of nuclear weapons-grade highly enriched uranium to the United States in Project Sapphire, Weber added, "Let's say that's fifty atomic bombs that cannot be made, so I felt pretty good [about that too]."[1]

On the Stepnogorsk project, Weber painted a larger picture than simply destroying the anthrax factory:

> The ability to [work with the Russians on securing pathogens]—which at the time I didn't realize how innovative it was—[was critical] to [detecting] Iran's efforts to acquire pathogens, technology, and scientific expertise from the Vektor laboratory in Siberia and the Engelhardt [the Russian Academy of Sciences's Engelhardt Institute of Molecular Biology] laboratory in Moscow. That was an incredible experience, because something really bad was happening, and traditional US government tools to stop it weren't working. We approached nonproliferation at the source, and with the directors of the institutes, the scientists working at the institutes.
>
> Essentially, we cut a deal, which was, "We can work together, world-class scientists, but you really have to break off your ties to Iran because we know that the Iranians you're in touch with . . . their primary interest is developing weapons of mass destruction." It was a tough negotiation, but it was at the source. . . . It was very much a direct relationship with the institutes.[2]

Sig Hecker distinguishes among four major areas of concern in the nonproliferation area: "loose nukes, loose materials, loose people, and loose exports." "So we look back now," he told me, "and what did we accomplish? Loose nukes? There weren't any. Loose materials? At the beginning, they had maybe 1.4 million kilograms combined of plutonium and highly enriched uranium, and there were these reports of the theft of some plutonium and you noticed maybe up to a kilogram some highly enriched uranium, but essentially over the years almost nothing. You know probably altogether, if you look at the IAEA [International Atomic Energy Agency] database, not one significant quantity *lost* altogether. Any loss is a concern, but it's just not there."[3]

What about "loose people?" There was only one such documented case (as both Sig Hecker and Laura Holgate confirmed in my interviews with them): that of Vyacheslav Danilenko, a scientist who had worked at Chelyabinsk-70 and had allegedly revealed secrets about nuclear weapons design to Iran.[4] Of Danilenko, Hecker commented:

> Yes, Danilenko, he presumably did some explosive-related stuff . . . with the Iranians. It is not clear exactly what impact that had. . . . But potentially it could have been disastrous—I mean a couple of their designers helping the North Koreans out, you know, would be just a disaster. It didn't happen. We explained that in the book,[5] and again, the Russians explained, they said,

"This wasn't going to happen; but we did have a crisis, we couldn't pay our people, our people lost and so you guys came along and helped and provided some help and provided some employment and provided some payment and provided—just the fact that we could interact together." . . .

Loose people? As I just mentioned, maybe one person did some things, and there were probably a few others; there was some activity related to laser enrichment and Iran. . . . It is that this just was their big patriotic duty [to be loyal and to stay put]. . . . And then the fourth problem: loose exports? Again, not much. Some went to Iran in the 1990s of concern, but if we look at exports today is just not that bad. . . .

So what did these cooperative programs accomplish?

You had the perfect nuclear storm and it is true that what was accomplished was the absence of things. Loose nukes, loose material, loose people, loose exports. But that is a major, major accomplishment. And today, in each of those areas, there are not loose nukes. They secured the material; that's what I spent most of my time on working with the Russians: the MPC&A—nuclear Materials Protection, Control and Accounting. I brought that into Russia to have them change from the old Russian days. And indeed, now, you don't dare steal anything from them because the consequences would be grave. So, those things didn't happen.[6]

To be sure, proving a negative is not logically possible. How to demonstrate conclusively that a person or persons had *not* gone to work for rogue states or terrorist organizations? But there were plenty of tangible, demonstrable achievements: Shipping six hundred kilograms of HEU from Kazakhstan to the United States and blending it through Project Sapphire, dismantling nuclear warheads, and upgrading security at the Russian nuclear labs and securing the nuclear material there —all this was an undeniable accomplishment of historic proportions.

What was the role of scientific cooperation in this historic effort? Was the nonproliferation program merely a bailout, a quick fix, a make-work program, essentially a bribe to keep people in place without any real substance? Not at all. The scientific cooperation was essential, according to Hecker:

The science that we did together, the science that we did—those four hundred papers—they were more important in opening the doors for nuclear weapons safety for nuclear material security. These things were all interlinked together so you can't separate out and say this one was successful, this one was not successful. The bottom line is you had this for loose-nuke problems and in the end those things didn't happen: That's success. . . . It was worth every penny of the Nunn-Lugar money and it was worth these thousands of lab people going back and forth working together, developing the relationships, doing those things that are difficult to do through the government.[7]

Regarding his work to prevent the spread of biological WMDs, Weber made a similar point. He told me:

> The peer-to-peer relationships between scientists were really the magic that made this happen, and one of the great accomplishments of that bench-level cooperation was the transparency provided. We really only wanted to know one thing. There was only one [thing] that we wanted to know in places like Vektor, and that was: Were they no longer working on biological weapons? And the only way we could do that wasn't through inspections or arms control or visits; it was with that day-to-day interaction between bench-level scientists. So I feel we were very successful in increased transparency and deep knowledge of what they were doing, what their funding was, and a side benefit of that—I call it a side benefit—of what was essentially a nonproliferation program, was that they produced some good research.[8]

While the lab-to-lab programs were very important in promoting joint research, the main vehicles for scientific cooperation with the former Soviet WMDs were the formal nonproliferation programs. Of all these, the International Science and Technology Center (ISTC) was the granddaddy of them all and by far the most extensive. Here is how Glenn Schweitzer, the ISTC's first executive director, summarized it in a personal memorandum:

> When the ISTC became operational in 1994, I thought we should plan for a lifetime of about ten years. It was clear that it would take a few years to energize the Russian S&T establishment, although sooner rather than later the Russian government would become tired of being subservient to the other countries that controlled the money. For a few years, the enthusiasm of the Russian research community clearly outweighed the lack of control of the government officials over developments entwined with the security and economic future of the country. But at the end of ten years, the writing on the wall became increasingly bright and indicated that it was time to either change the way the ISTC operated—for example, having a Russian as a cochair of the governing board—or prepare for a prompt exodus and transition to a new form of S&T collaboration.
>
> That said, overall, the ISTC was a success story of the first order. The number of successful projects in advancing science was remarkable—and at a relatively low cost. The opening of the closed cities throughout the country defied expectations as the Russians reevaluated the need for keeping so many activities under classified wraps. Projects such as the Mission to Mars; biosecurity/biosafety; nuclear research, including cooperation with CERN; and design and testing of aircraft innovations were among the many areas of contributions to global S&T interests. Now, many years later, the equipment provided by the ISTC is still playing a critical role in operations at many laboratories throughout the country. The cross-ocean contacts made through the ISTC spawned many enduring relationships. The commercialization activities of the ISTC may have sputtered in the beginning but within a decade, they

changed the mentality of many important Russian scientists and transformed the concept of innovation into the reality that innovation is only innovation when it reaches the market place. While the senior Russian scientists who led most of the projects may be retiring, many of the early career scientists during the days of the ISTC are now still making their marks throughout the Russian research community.[9]

Hecker agreed: "It's not fair to separate the people programs . . . from the nuclear material safety thing, because many of those people were the same people. They were deeply connected and the good that you did with the people program built what I call a bank account of goodwill."[10] But it was more than a bank of goodwill that was at work here. It was professional respect and personal trust built on the foundation of high-quality scientific research in some very sophisticated areas that were unique to the two scientific communities. It was the level of mutual confidence bred by that research that enabled not only the scientists but also their governments to address the towering practical issues of nonproliferation of WMDs in an effective manner.

A 2013 assessment by the National Academy of Sciences on the various US-Russia bilateral programs to address the proliferation of biological WMDs reaches similar conclusions, but it is more guarded about the actual scientific outcomes of the many collaborations that took place under their auspices.[11] "In the area of national security," it states,

> U.S. financial support during the 1990s and early 2000s of Russian endeavors to enhance biosecurity and biosafety approaches and capabilities substantially reduced the risks associated with possible misuse by malcontents of the biological assets of Russia. As an important component of this effort, the United States joined with Russia in supporting redirection of thousands of underemployed Russian scientists in the defense sector to jobs in the civilian sector that provided pay supplements during economic downturns in the country. The joint activities have also upgraded the equipment bases and related infrastructure weaknesses of Russian research institutions. . . . And at times, the programs have responded in a modest way to the Russian government's near-time priorities for development of saleable products and services.

With regard to the actual scientific outcomes in terms of the development of new knowledge, however, the report concludes that after some twenty years, the jury is still out: "Scientific advances that can be attributed at least in part to cooperation are still unfolding. But some progress seems clear. . . . Of most importance, new international networks among scientists have been established and are being maintained. . . . A number of Russian-authored and coauthored international journal articles can be attributed in part to bilateral scientific engagement."[12]

Not everyone, however, judged the nonproliferation programs to be an unqualified success, even in terms of the goal of nonproliferation. I had an interesting conversation with Laura Holgate that brought out this point. My interview questions asked the discussant to evaluate his or her actual accomplishments and experience in light of their initial expectations. Holgate took that literally. Being careful to frame her answer in terms of the initial thinking behind them—that is, to prevent the former Soviet WMD scientists from continuing to develop WMDs in Russia or to disseminate the weapons or technologies to rogue states—she responded that the programs' effect was "a drop in the bucket."

"A drop in the bucket?" I asked with some surprise. She went on to clarify that the programs' success, as initially conceived, was dependent on the health of the Russian economy and, in particular, a "normal market." "The premise," she said, was "that there could be an economic driver or pickup" to build on the US funding, essentially a "subsidy," to make the nonweapons outcome sustainable independent of external funding. However, since, as she pointed out, the twin threats from rogue states (countries) and internally, from the former Soviet weapons labs, never really materialized, the basis of the scientist engagement component was not as solid as had been thought. Regarding the scientists themselves, she remarked, "We totally didn't get it." That is, the United States (and other Western countries) never imagined that the former Soviet nuclear scientists were in reality "homebodies," seeing themselves first and foremost as loyal citizens and indeed patriots. "We totally misjudged them," she continued, about "whether they could be bought" or "whether they'd go somewhere else." In the end, however, as regards the scientists, she concluded, "I don't think any harm was done—some good was done."[13]

In this assessment of the likelihood of Russian WMD scientists selling their work to the highest bidder or running to other countries in search of lucrative jobs, the common wisdom was, in fact, dead wrong. Indeed, in her 2002 scholarly article on the "brain drain" phenomenon, Russian science and technology policy Irina Dezhina wrote at length about the "creation of myths on the question of 'brain drain.'" Due to inadequate statistics and sensationalist reporting, she wrote, "the extent to which this subject spread like weeds can be compared even with a computer virus."[14]

However, for present purposes in considering the impact of the nonproliferation science cooperation programs, it is important to keep in mind that the original motivation for the immense and costly programs of the United States and other countries in the area of nonproliferation was based precisely on the fear of this brain drain, especially from the former Soviet institutes and facilities that researched and produced deadly nuclear, biological, and chemical WMDs. The hemorrhage imagined by policymakers and pundits alike never occurred, and in fact, according to our witnesses, was never about to occur. One could argue that without these programs it might have happened, but again, we confront the impossible challenge of proving a negative.

What we do know is that firsthand observers who knew these people intimately told us, over and over again, that it was not going to happen. The real problem, as Hecker and Holgate and many others who went over to Russia and elsewhere to visit these facilities have pointed out, was not what Hecker calls the "loose people" but the "loose material"—highly enriched weapons-grade uranium sitting around on factory floors and storage facilities with loose padlocks and drunk security guards or dangerous pathogen collections on shelves in relatively open scientific institutes. Once this was discovered, the US government undertook major programs to remedy these real dangers, and by all reports they were quite successful. And as Hecker and Weber correctly noted, the professional respect and trust built up through the personal ties born of cooperation on highly specialized questions of basic scientific research enabled both sides to go on to solve these more difficult problems.

Nevertheless, the myth of the much-feared "brain drain" had lasting consequences that were to haunt the relationship later on. Not only was this perception wrong—or, let's say, misguided and misinformed—but as the founding myth of some of the most extensive and expensive cooperative science programs ever known, it became the basis for what I increasingly saw as a warped, condescending, and enduring perception of the problem, and most of all of the long-term impact on the people most involved in the problem—the FSU scientists themselves.

The frame of reference of these people-oriented nonproliferation programs—the goal of which was variously characterized by their managers as "scientist redirection" or later, when that term went out of vogue, "scientist engagement"—was rooted above all in the mode of unilateral assistance characteristic of the early post-1991 period of emergency financial support. As I kept trying to understand it, the thinking seems to have gone something like this: If we were to provide these WMD scientists with sufficient material incentives to leave defense work entirely by paying them to work on nondefense projects, eventually they would find a way to use their professional backgrounds in sustainable civilian pursuits.

But as Holgate pointed out, any such expectation was based on the premise of a serious market for the resulting technologies—let alone the scientists' capacities to innovate in a market environment. These were not sound expectations in the first place, and in any event it did not happen, as was obvious early on.

The operating mode of these programs, moreover, reinforced this kind of wishful thinking. They provided unilateral support for work going on in Russia, but little if any support or infrastructure to promote genuinely cooperative projects with Western scientists. (The fact that there were "partner programs" that supported genuinely cooperative work is not helpful here, since as I have pointed out in chap. 4, the partner programs were essentially flow-through activities that did not involve program funds and did not significantly change the programs' overall goals or approach.)

But with time—and not very much time, I would (and did) argue—this kind of thinking became obsolescent. It was not going to be assistance, no matter on how grand a scale, that would provide a sustainable path for these fine scientists and engineers toward self-sufficiency in "peaceful," nondefense work, for three reasons. First, assistance by its very nature is a temporary entity. Policymakers will continue it only so long as it can be conclusively demonstrated that it is needed, and that it has worked. As emergencies fade from the scene and get replaced with new exigencies, this is always an increasing challenge. Second, as argued above, the domestic market of any conceivable resulting innovations simply did not exist. The new wealth in Russia was far more interested in luxury goods and foreign high-tech products, and Russia continued to rely heavily on extractive industries rather than developing a knowledge economy. Most importantly in my view was the nature of the incentive for the scientists themselves. As we have seen in chapter 6, probably the most powerful incentive for formerly isolated Soviet scientists was this: a path to joining the world scientific community. But alas, the nonproliferation programs never really offered that. And that is the pity of it.

I witnessed this logic play out firsthand with the Global Initiatives for Proliferation Prevention (GIPP) program when I worked at the US Industry

Coalition from 2006 to 2012, the final years not only of the organization but also, as it turned out, of my remunerative career. To the very end, the National Nuclear Security Administration (NNSA) managers were unable to wean themselves from the notion that a meager per diem of thirty-five dollars was adequate to keep the Russian nuclear weapons scientists engaged and happy, and they became defensive and combative whenever some skeptical or hostile member of Congress suggested that the Russians were in fact still working in the weapons institutes on weapons programs and that actually the nonproliferation programs were subsidizing the Russian nuclear weapons programs. Whether the latter allegation was true or false I cannot say, but I suspect that the truth was somewhere in between. In any event, the nonproliferation programs in general continued to operate largely in the mode of "assistance" and were never able to get past that kind of thinking and structure, despite calls late in the game, well into Vladimir Putin's presidency, to shift to a more cooperative, truly bilateral, framework.

Glenn Schweitzer relates how he sensed in 2009, on the occasion of a celebration in Moscow of the ISTC's fifteenth anniversary, that the atmosphere was souring. "Sensing the push for dramatic changes in the ISTC's future ranging from closure . . . to its replacement by an organization that focused on science cooperation rather than proliferation . . . I proposed several adjustments to the political and policy framework of the ISTC." These included appointing a Russian co-chair to the governing board; inclusion of Brazil, China and India as nonvoting members; increased attention to counter-terrorism issues; and better metrics. "I regret that none of these global suggestions were seriously considered. They simply were offered too late in the lifetime of the ISTC."[15]

Eileen Malloy, who spent two years in the NNSA front office overseeing nonproliferation programs, cites a relatively mundane example of this type of thinking involving what would appear to be a purely logistical matter: accommodations for visiting foreign delegations.

> The difficulty with assistance . . . is that if you categorize it as assistance you then have this enormous trail of GAO [Government Accountability Office] accountability. And that was the bugbear. . . . What happened is that people were doing things to assist the scientists but not calling it assistance. . . .
> That was a huge problem that I had when I got to DOE, because they had deformed the bilateral process on the ground. And by that I mean, if we wanted to go to a nuclear city, to go on site, and have talks, there were no hotels there . . . no Marriott. So the only place you could stay was in a guesthouse facility there. The Russian scientists did not have the money to keep those

up, and they did not have the money to take care of visitors. I get that. So we needed to pay our way. But at some point, somebody in DOE decided, well, we'll just pay them at the world rates for lodging. So they were paying, you know, one hundred plus dollars a night to stay in something that wouldn't get a one-star rating.

I'll give you an example. I took a delegation there once and the deputy secretary [of energy] was going to spend the night in a room that didn't have a bathroom, and so they were going to put a little chemical toilet in the corner and asked, "Is that OK?" No. It's not OK! But what I said to my fellow DOE staffers was that we should only have been paying them what they needed to host us. So whatever it was in the beginning when they had absolutely no money, wouldn't it have been far better for the local entity—the Russian lab—to contribute that as their contribution, so that the cooperation would not appear to be so one-sided? And what this did is that they became so dependent on us paying unreasonable rates for things. It became a money cow.... The Russians thought that we were just coughing up this money so that we could get in and see things that we wanted to see. And the fact that we were paying these outrageous sums just fed the paranoia on the part of the Russian secret services.[16]

This was a problem that went way beyond guest accommodations and toilets. It had to do with the controversial role of the nonproliferation programs in providing indirect-cost support for infrastructure. As anyone familiar with research grants knows, the support of institutional costs attributable to a research project but not direct research costs themselves—"indirect costs"—are an essential part of the landscape of all grant-making programs. While there was often vigorous debate about their scale and methods of calculation, it is widely understood that research institutions could not survive without indirect costs. The new US-Russia research grant-based programs of the post-1991 period followed suit. They typically included a flat "indirect cost" fee of 10 percent that was meant to cover management and operational expenses that could not otherwise be attributed directly to the research projects. This was true of both the purely civilian grant programs, such as CRDF's, as well as of the nonproliferation programs. In the latter, however, indirect costs paid to the Russian researchers' home institutes actually became a hot issue for GAO investigations and congressional hearings, because critics accused the programs—especially in the case of GIPP—of unwittingly subsidizing the Russian weapons programs because the indirect cost funds paid to the closed institutes were essentially unaccountable.

By the time of about 2010, the "scienctist engagement" programs began to encounter another vexing problem, one of hard demographic reality. The "former WMD scientists" were dying out or retiring and there were fewer and fewer people, and indeed institutes, that could plausibly be considered in the "former WMD" category. Thus, in response to congressional pressures, the ISTC began "graduating" Russian institutes from the program, while the GIPP program cast a more discerning eye on its FSU client scientists at the individual level. These last-minute measures, however, did little to assuage congressional skepticism, despite the agencies' heroic efforts in repeated hearings and in response to GAO reports.[17]

None of this is to say that the cooperative science programs in the nonproliferation dossier were bad or failures or helped to aid and abet the adversary, as some of the most virulent congressional critics were saying in 2007 and later. I believe that the reverse is true: they were well intentioned, well designed, well funded, managed by knowledgeable and able people, and had many beneficial results, both scientifically and in terms of global security. This cannot be overemphasized. All the same, in the end they failed to adapt to changing times. One of the key reasons they did not adapt is that the underlying concepts and goals of the programs, while understandable enough and praiseworthy at the time of their origin, became idées fixes in the minds of their advocates, defenders, managers, and even beneficiaries in both countries, long after those concepts were reasonable or appropriate. In the end, the programs increasingly became seen by Russians as unilateral intrusions into their internal affairs.

Beginning in about 2010, access to the closed facilities became increasingly difficult. The Russians were signaling "that the redirection of Russian weapons scientists to civilian endeavors had been completed."[18] Finally, in August 2010, the Russian government gave formal notice of its intention to withdraw from the ISTC, and a year later finalized its withdrawal to coincide with the termination of the last ISTC-funded project in 2015.[19] More or less in parallel, the GIPP's final congressional appropriation expired with the passing of fiscal year 2014 (ending September 30, 2014).

In January 2015, the distinguished former Senators Sam Nunn and Richard Lugar, the creators of the visionary "Nunn-Lugar" nonproliferation programs named after them in the early 1990s, wrote an eloquent op-ed in *The New York Times* lamenting that the Congress, a month before, had voted to cut off funding for the last surviving bilateral US-Russia nuclear security programs in the wake of the Russian annexation of Crimea. "We need a new approach," they wrote, "a real nuclear security partnership guided by the principles of reciprocity and mutual interest, to which both countries contribute their own funding and technical resources."[20] But it was not the

Russian aggression in Ukraine that killed those programs. It was, instead, the failure of the programs, for many years, to articulate and act in the manner in which Nunn and Lugar suggested in their op-ed, whether due to institutional inertia or lack of imagination or will, that had long before doomed these programs to their demise. The 2015 Nunn-Lugar op-ed was, as the saying goes, a day late and a dollar short.

Promoting Foreign Policy Goals

I have always had a bit of trouble with finding concrete examples of how bilateral US-FSU cooperative science programs promoted US foreign policy goals. It may help to divide this category into two parts: symbolic goals and real ones.

Symbolically, the cooperative science programs of the US government have sometimes been good, if convenient and not costly, ways of expressing disapproval of unfriendly or repugnant actions of the Soviet or Russian governments. In the 1970s and 1980s, it became commonplace to get instructions from the State Department or the National Security Council that there were to be no "high-level visits" or meetings in response to one Soviet outrage or another—the treatment of Andrey Sakharov, the invasion of Afghanistan, the imposition of martial law in Poland, the shooting down of a Korean airliner, and other lesser-known incidents. Sometimes, these situations led to the cancelation of entire programs—but never all of them at once. When asked about the rationale for these actions, the usual answer from the State Department was "we're sending a message." Program managers like me began to tire of the use of our cooperative programs, which we believed to be meritorious in their own rights, as essentially carrier pigeons for diplomats and others who seemed to have difficulty finding their own voice to "send messages." To the people who used the cooperative science programs to send messages, the programs were pieces on the chessboard of international diplomacy and little else.

Moreover, in no single case that I can remember did the "message" ever result in any perceptible change in behavior by the other side in its offensive behavior. And this is where I reserve a special place for Secretary of State Henry Kissinger's congressional testimony of 1972, when he characterized US-Soviet science cooperation agreements as "incentives for restraint."[21] Over my many years in this field, I failed to find a single example of that actually happening. The Soviets did not relent on their persecution of Andrey Sakharov or Jewish scientists because we canceled a working group meeting in physics or some other area; they did not pull out of Afghanistan even when we boycotted the 1980 Olympics; they did not rescind martial law in Poland. True, when things started getting better, as during perestroika, the US government put some agreements back in place, such as the basic sciences agreement I helped to negotiate in the late 1980s, ramped up funding for some programs, and in general returned to status quo ante. But to see these steps as "rewards" for good behavior, as I actually have heard many diplomats and policymakers say over the years, was really too much to take.

But there were real casualties of these punitive, symbolic actions. The casualties were the scientists in the former Soviet Union (FSU) who had cast their lot into cooperation with the United States, for whom such cooperation was also a stepping-stone to membership in the world scientific community. Any Soviet scientist who engaged in these programs had to be vetted and, as is well known, occasionally visited by members of the state "security organs" to verify their reliability and, if possible, to serve as information conduits. And when these same programs were suspended or canceled by the United States in protest, life undoubtedly became less comfortable for these people. By the same token, when programs were allowed to resume as a "reward" for good Soviet behavior, among those who were being rewarded were in fact members of the US scientific community whose scientific research was enriched in various ways by access to people, institutes, and sites in the other country.

Of Henry Kissinger's dictum, I have often wondered whether he actually meant it when he told a skeptical Congress that the bilateral science agreements were intended as "incentives for restraint." In fact, as we have heard from Norman Neureiter, when Kissinger came to his boss in the White House Science Office, Presidential Science Advisor Edward David, in 1971, he characterized his purpose to David as wanting "to offer the Chinese something more than the geopolitical repositioning that would occur—something more tangible, more concrete; perhaps, he said, some proposals for science cooperation that will show the Chinese we are serious about some kind of enduring engagement."[22] In other words, he was sending a message. He knew very well that the proposals were in no way going to be in themselves "incentives for restraint." Evidence of good US government intentions, perhaps, or an overture to establish a closer relationship that might, in other ways perhaps, moderate Chinese behavior, but science projects as incentives for restraint? I don't think so.

Perhaps I am being too literal here. Thomas Pickering, one of the deans of the US diplomatic service with deep, unrivaled experience and understanding of world affairs, a former ambassador to Russia and a host of other countries, and an ardent supporter of international scientific cooperation for decades, took a longer view when we talked:

> Well... the bigger incentive for restraint was essentially the notion that we had to be very careful not to end up in a nuclear disaster. And that arms control and disarmament... was totally informed by science. [Arms control treaties] couldn't be operated without a clear indication and interest in how delivery vehicles and weapons work. [Scientific knowledge and understanding were] critical in formulating the rules for limitations as well as even, more importantly, the capacities for inspection and national technical means—things of that sort. Secondly, the joint investment in things that became a win-win and had a high priority in terms of national interest is in itself a serious restraint upon taking kind of mindless, negative actions. And so to some extent, Putin is perhaps in part the exception that proves that rule. But even he, I think, is limited.
>
> And I've always thought we should take an intensive and careful look at our relationships with countries that I call "rivals and partners"—China, Russia, India; probably Japan and Brazil; and you can find one or two others—and that our relationships with them, particularly at the high end of contention possibilities, need to be based on working hard to find those areas of salient common interest in which we need to develop a set of relationships that can act as a brake on the negatives, that can become a framework in which you can deal with the negatives, that permits you at a minimum, with respect to the negatives, to do what I call "Hippocratic diplomacy"—to stay away from doing harm. But then you can isolate and compartmentalize them, but then see if you can chip away at finding answers.
>
> The alternative to that, on the other side, at the other end of the continuum, is for those who particularly are single-interest oriented—on proliferation, on human rights, whatever—to overemphasize the negative issue and try to use sanctions and pressure to resolve it and as a result, on both sides, to create a deeper sense of antagonism, uncertainty, nervousness and bad action. And somewhere in the continuum in between, there may be a balance, but my view is that it is much better to look at what I would call looking for win-wins and using that as a defining characteristic of the relationship. And it's exactly what Henry meant by saying using science as a restraint. So science in itself may be in some areas even so important in itself to be ipso facto that restraint, but it informs much of what else has to be done purely on people with no scientific background, thinking about how to develop the foreign policy relationship and screwing it up.[23]

It's hard to argue with that statement, especially considering its source. But I am still skeptical that that's exactly what Kissinger meant when he talked about "incentives for restraint." In the context of his congressional testimony, it looks more like a smokescreen than a statement of fact or even intent. In any case, this concept underlies a highly instrumental view of scientific and other types of cooperation not only as tools or instruments of broader policy engagement, which they certainly are, but also as reified pawns to be manipulated, used, and discarded, as a substitute for direct speech or action. Anyway, that's why I never became a diplomat, and why Kissinger is thought of as the world's leading living example of realpolitik.

Commercial Accomplishments

In the formal bilateral cooperative science programs, achieving commercial results was never an end in itself. But it could be a metric of success of a certain kind. Moreover, the commercial metric certainly did have some appeal to members of Congress and the public who called for concrete evidence of useful results and benefit to the United States from taxpayer-funded activities which, as was a common public perception, was more to "their" benefit than to "ours." This was particularly common beginning in the 1990s, when the overall rationale for publicly funded cooperative programs with Russia and other FSU countries morphed from mutual benefit into assistance. "Assistance" was not in itself enough to satisfy these concerns. In response, some agencies attempted, with greater and lesser success—mostly lesser— to conjure up some modes of activity that appeared to have commercial benefit. In addition, of course, outside of the programmatic framework, for-profit companies, both large and small, sought to capitalize beginning in the 1990s on the availability of the sudden availability of highly qualified technical talent at bargain-basement prices, as well as the opportunity to develop vast new markets.

I did not personally have much exposure in the earlier part of my career to the purely business-to-business side of science and technology cooperation with the Soviet Union or its successor states. From the Soviet period, one example I was aware of was in the area of advanced welding techniques. The Paton Institute of Electric Welding of the Ukrainian Academy of Sciences in Kyiv performed world-class basic and applied research in welding, including electroslag welding, electroslag remelting, metallurgical coatings using physical vapor deposition, and flash-butt welding. Their works' applications ranged from tanks and submarines to construction, medicine (welding of tissues), and more—including building pipelines.

In 1979, when I arrived at the National Science Foundation, one of the world's largest construction projects was the building of the Alaskan pipeline. Pipelines were nothing new to Western oil and gas companies, but pipelines in frigid, cold-weather conditions were a new challenge. It was apparently common knowledge that to solve this problem, they purchased a license to some of the welding techniques developed by the Paton Institute. And as it turned out, these welds have stood the test of time. When I was in Kyiv in December, I ran into an oil man from the Alaskan North Slope whose job was to perform quality control on the people and hardware of the massive operation there. (He was in Kyiv for personal business.) I mentioned the Ukrainian welds to him, and he told me that of all the components of the pipeline, the welds have held up the best—"flawlessly," he said. As in the case of missile silo technology,[24] Soviet experience in cold-weather conditions was a perfect fit for the commercially driven work in Alaska. That example, and the early work of John Kiser on what he called "reverse technology transfer," alerted me that the common wisdom that the Soviets had little to offer the United States that was of a commercially useful nature was not quite right.[25]

The first insight on this question from my interviews comes from Thomas Pickering, reflecting on his post-ambassadorial work at Boeing. In this case, the cooperation was purely of a business-to-business nature and mostly in engineering, although advanced titanium metallurgy was also involved:

> When we built the 777 at Boeing, we had a choice between a landing gear of four four-wheel landing gears or two six-wheel landing gears. The Russian TU 0154 had a six-wheel landing gear. So we worked on the six-wheel landing gear. We worked with the Russians on testing it. And we continued to buy titanium, and that was huge and it still is. And the largest single item they provide us in titanium was the main beam that linked up those six wheels on each side—the truck beam, it's called—and when we started out, they did that for us as a forging and then we machined it in the States and it became very clear partway through the operation that we could do the machining in Russia in a joint venture, connected with the titanium production, much cheaper and much more effectively than we could do by shipping big parts to the States. And so we in a sense offshored more of the technological basis of that work....
>
> There was a guy who ran the titanium factory whose name was Tetyukhin. He's a metallurgist and he did metallurgical development for us. He developed ... titanium alloys for the aviation industry, which was particularly useful to Boeing. So, Boeing has very large business on a continuing basis, and of course we're very concerned that politics not get in the way of that. I mean, we're fully backed up now with alternative sources of titanium, but the Russian titanium in our operation has been very mutually beneficial for both of us. And in the course of this we have sold many hundreds of airplanes into Russia. They helped to make them, so they helped to buy them.[26]

Boeing set up a large research facility in Russia, as did Microsoft, Intel, and other US and multinational corporations. In the information technology area, the main source of interest was Russian expertise in software programming and design, but as Sig Hecker reminded us, the Soviets actually had parallel processing before the United States did. These were purely corporate decisions, unrelated to any particular government program.

In the government-sponsored or funded collaborative programs themselves, development of technologies for practical, commercial application was often something of a holy grail, largely for political reasons as argued previously. Commercial benefit was something that the American public and their elected representatives could understand and grasp, even if it was not the main purpose of the activity. In theory, it was concrete and measurable. And of course, for scientists in the FSU who desperately wanted to learn how to turn their technological skills into wealth and sustainable jobs, there was great eagerness to engage in such efforts and learn the ropes.

There were intrinsic problems, however, with the notion of government programs promoting commercial development. Traditionally, aside from contractual relationships and specially targeted programs such as the Small Business Innovation Research and Small Business Technology Transfer programs, the US government keeps an arm's length from funding private companies to do research. The exact distance varies over time, largely based on which political party controls the White House (and, if possible, the Congress). Democratic administrations have traditionally been more attracted to the idea of targeting funds even to private companies for innovations that are particularly promising in terms of potential social benefit, such as in energy and environment, while Republican administrations tend to decry "picking winners" in the private sector.

But more deeply, it is largely an article of faith in government agencies that they do not give away money to the private sector. So when it came to using taxpayer funds for programs that encouraged private technology commercialization, even if those programs were intended to promote public goals, such as combatting the proliferation of dangerous technologies, the government encountered a gray area.

The nonproliferation programs were a case in point. While the ISTC, for example, aspired to the goal of achieving commercial results from its scientific engagement projects, "there is little evidence," wrote Glenn Schweitzer, "that this program and related efforts had a major impact on moving research results into the market in the short term." He attributed this failure to "the stifling effect of rampant corruption, lack of transparency, and a cumbersome Russian bureaucracy on the efforts of small entrepreneurs to penetrate significantly into the Russian market."[27] These reasons are certainly true, but two other factors played an equally important role: the inability of US government programs to effectively leverage such opportunities because of ideological and other obstacles and Russia's lacking market for high-technology industrial products, a far more serious problem. This is a very broad topic that is beyond the scope of this book, but astute observers have often remarked that the Russian government's economic strategy of relying primarily on natural commodities such as oil, gas, and wood, combined with an extremely lopsided distribution of wealth in the new Russia that favors luxury consumer goods and not industrial development, have been very serious inherent obstacles to the commercialization of knowledge-intensive technologies in the Russian domestic market in general.

This is not to say that the ISTC functioned in a complete commercial vacuum, either. Because of its special status as an international organization, giving it important tax and other privileges, the ISTC—as well as other organizations, such as the related Science and Technology Center Ukraine (STCU) and CRDF Global—was able to handle financial and administrative transactions for both for-profit companies promoting scientific research in the FSU (in the case of ISTC and the STCU, only with scientists with a weapons background) and to thus be an indirect enabler of commercially important activity through its good offices. Nearly one-third of its funding in its heyday came from so-called partner projects. But it was not the development of commercial technologies that the ISTC sought to achieve through its own activities, despite its valiant attempts to do so.

The GIPP program of the Department of Energy's (DOE) NNSA was more integrally designed to include commercial participation and success as a core concept and process. As related in an earlier chapter, the typical GIPP project consisted of an FSU weapons lab, a US DOE national lab, and a private US company that was called an "industry partner." NNSA directly funded the work of the FSU lab, funded a portion of the work of the US national labs (but never enough, in their view), and gave a small subvention each year to the US Industry Coalition to manage the nonprofit industry association of US industry partners. The industry partners, on their part, were required to dedicate an amount of money, in cash or kind, equal to the investment of NNSA on the FSU side of each project. The US labs owned any intellectual property that was created during the course of these projects and granted unrestricted licenses for its use to the FSU scientists as well as to the US industry partners. It was a very complicated framework with a lot of moving parts and business arrangements—not the kind of thing that was easy to explain to members of Congress or the public. But, to some extent, it did work.

Three individuals who participated in either the GIPP or CRDF programs, or both, were Terry Lowe, Randolph Guschl, and David Bell. We have met each of them earlier in chapter 5, when they told us about their paths to becoming involved in Russia and Ukraine. Here is what they had to say about the results.

Terry Lowe had left Los Alamos in 2000 to work in the private sector in the blossoming area of nanotechnology. In 2001, he cofounded his own company, Metallicum, and through it became an industry partner in a nanomaterials project with Ruslan Valiev of the Ufa State Aviation Institute. Valiev and his group were very impressive and innovative, as described earlier in chapter 7. Together with Lowe, they developed promising results, and Lowe attracted major investment from an entrepreneur, Marvin Maslow of Manhattan Scientifics. Lowe said:

> We . . . sold Metallicum to Manhattan Scientifics in 2008, and it became a Manhattan Scientifics subsidiary. . . . We were successful and that was largely because of the folks at Manhattan Scientifics taking the technology and introducing it into the mainstream metals manufacturing community. . . . They adopted the technology. We didn't seek them out; they actually sought us out. . . . During that period of time . . . Manhattan made many millions of dollars. . . . Now there are other metal manufacturers . . . navigating into a sensitive area, I would say, that are looking to adapt the technology and multiple companies that are in product trials right now, to look at putting it into their products. So the technology works. It is a good, profitable company, and actually all of its profits have come from this technology, but I profit from it too! . . .
>
> And incidentally, my relationship, and the relationship with the Russians, has continued throughout this period. We have actually done joint commercial projects together independently of [GIPP]. . . . And we're looking at other areas; we've actually proposed other joint projects recently as well. So it's just continued to grow, and it goes in different areas. We worked on titanium; now we address magnesium and [its alloys] and its applications to steels, stainless steel in particular . . . , so it's just continued to blossom, and it will continue to do that in the future as well. To a certain extent, Valiev and the group are aligned with us, but in other areas, they have diverged and gone off in other directions, which you expect, of course. The level of commercial success they have compared to what Metallicum has done is [an order of] magnitude different. Their success is quite small; they founded a company called Nanomet, and it has in fact been producing and selling nanostructured materials, but the volume that they do is measured in the hundreds of thousands of dollars; there's nothing

north of ten million dollars. We're in the ten-million- to one-hundred-million-dollar range, and they're in the hundred-thousand-dollar range.[28]

This was certainly one of the most commercially successful projects to emerge from the GIPP nonproliferation program. Not only did it "engage" talented Russian scientists, but it also in fact resulted in sustainable livelihoods for them. It was the Russians' unique research skills and technical capabilities that primed the pump.

In another case, it was not so much scientific skills as much as access to unique physical collections that made the difference. When I asked Randy Guschl of DuPont about his company's outcomes through the GIPP program, here is what he said:

> Well, I can't answer that because the success of R&D is ten years later, when you look at what's come out of it.... But I do believe that the number of leads that our plant science people pursued as a result of dealing with a large number of samples from libraries has got to be considered a success, because Pioneer on its own went back with a lot of their own money. They wouldn't go back if they didn't feel it was there....
>
> So I would say that's the single biggest success.... Another area was general catalysis, and also some work related to fluorinated materials.... That was a rare case where there was a group focusing on organifluorian chemistry. Organifluorian chemistry is not done at very many places in the world. There was some good research there. And once again, the measure of success is when people go beyond the [external] grant money to spend their own money. That's success.... You heard me speak about this [at the 2011 USIC Annual Meeting], I think I said that cumulatively—we kept track of the government grants from the three agencies—it was about nine million dollars. And without being specific... there was about half again that much of DuPont's money that was following those leads.[29]

As a former competitive grant manager, I appreciated those remarks. To my mind, one of the measures of success of a research grant, aside from the publications and possible patents, was always a track record of competing successfully for more grants.

For Dave Bell of Phygen, the proof in the pudding was the "Miracle." Continuing the story of his work with Yakov Krosik of Israel, to whom he had been introduced by Russian Academy of Sciences member (now vice president) Gennadiy Mesyats, whom in turn he had met through the CRDF Next Steps to the Market program, Bell explained:

> In the course of doing that, Yakov introduced me to the Institute of Physics in Kiev, which was... teamworking on something that I found fascinating. That was a form of technology called "plasma acceleration." Plasma acceleration

offered the potential to solve the biggest problem that the coating industry had, in depositing better-quality films. It was very high-risk, maybe totally foolish on my part, but I end up bringing a scientist from Ukraine to work on scientific development here in this building in Minneapolis, and we did that for three years. . . . I did it with my own money and a very limited amount of investment money that I raised. We ended up perfecting that technology, plasma acceleration, and it is this coating that is the gold standard today for metal forming, dye casting, plastic injection molding, and so on.

They call it the "Miracle." We can take manufacturing components that might . . . wear out in eight or sixteen hours . . . and now we can make them last nine months. The "Miracle." So our company has been working to bring this technology to commercial markets where there's a need to enhance the performance of critical mechanical and wear components used in a variety of process applications and industries.[30]

The Miracle: a byproduct of scientific cooperation between the United States and Ukraine. It's hard to beat that.

Notes

1. Interview with Andrew Weber, April 25, 2016.
2. Ibid.
3. Interview with Siegfried Hecker, January 28, 2016.
4. See Gorwitz 2011.
5. Hecker is referring to the book Hecker, ed., 2016.
6. Hecker interview.
7. Ibid.
8. Weber interview.
9. Memorandum from Glenn Schweitzer, May 28, 2016.
10. Hecker interview.
11. *The Unique U.S.-Russian Relationship in Biological Sciences and Biotechnology: Recent Experience and Future Directions.*
12. Ibid.
13. Interview with Laura Holgate, October 13, 2015.
14. Dezhina 2002, 18.
15. Schweitzer 2013, 66.
16. Interview with Eileen Malloy, January 6, 2016.
17. See especially *Nuclear Nonproliferation: DOE's Program to Assist Weapons Scientists in Russia and Other Countries Needs to Be Reassessed.*
18. Schweitzer 2013, 64.
19. The text of President Dmitry Medvedev's decree of August 11, 2010, and of the Diplomatic Note on Russian ISTC Withdrawal of July 23, 2011, can be found as appendixes in Schweitzer 2013, 262–263.
20. Nunn and Lugar 2015.
21. See the discussion of Kissinger's view at the beginning of chapter 2. The quote is from Ailes and Pardee 1986, 11.

22. Interview with Norman Neureiter, January 11, 2016.
23. Interview with Thomas Pickering, September 24, 2015.
24. See chapter 7, "John Zimmerman: Missile Silo Technology."
25. See chapter 5, "The Entrepreneurs."
26. Pickering interview.
27. Schweitzer 2013, 46–47.
28. Interview with Terry Lowe, March 26, 2016.
29. Interview with Randy Guschl, March 21, 2016.
30. Interview with David Bell, January 14, 2016.

9

PROBLEMS

Despite the many impressive accomplishments of my respondents, not everything, of course, went smoothly. Here, we will look at specific types of problems that were mentioned in the interviews. I should also acknowledge here that most of the issues discussed below have to do with problems arising from the US side and not so often from the side of the former Soviet Union (FSU). This is an artifact, in part, of the geographic distribution pattern of my interviews, and partly also, perhaps, that the American interviewees (as well as their Russian émigré counterparts) may have been more inclined to be open with me and to feel free to identify obstacles arising from the American side.

At the same time, that testimony that I did receive from my former Soviet respondents, illuminates in graphic ways their direct experience with issues arising from within their own country. During some interviews, it was at times hard to keep my jaw from dropping at their candor. While readers familiar with conditions in the FSU will not be surprised to see such issues mentioned here, one does not often see them in print as narrated in the first person.

The order in which I present these issues below is not intended to be suggestive of the order of their importance or frequency. Informed readers may have their own preferences about which problems loomed larger or smaller; in this chapter, the choice was based largely on the frequency of what I heard from my interlocutors.

Inflexibility

All institutions and their activities have both advantages and disadvantages. The advantages are usually that they help people do collectively what they cannot do individually. The disadvantages are that they often get in the way precisely because of why they are often successful: the routines and processes they have developed to address different kinds of problems. These routines and processes can become reified, ossified, and counterproductive when they become seen as ends in themselves, or as the "tried and true" patterns that must always be followed, instead of means to achieve the ends for which the institution was established.

In the course of the scientific cooperation between the United States and the FSU, there are many examples of how institutions and arrangements meant to facilitate the goal of scientific cooperation actually had, or were seen as having, the reverse effect. The first time we encountered this problem was with the very first exchange programs—the IREX and interacademy programs.[1] Participants complained that their quota systems were artificial and rigid; and the two sides were constantly at loggerheads over how many scholars, and how many scientists, and in what fields, were to be apportioned to these quotas. One of the main reasons these cumbersome arrangements were put in place, of course, was a desire for control on the part of both governments.

At the same time, by codifying procedures and practices to manage the coming and going of scientists, these structures enabled their very movement to one another's countries. Thus, in addition to being instruments of control, the agreements and programs were protective and enabling "roofs" that made the exchanges possible in the first place. It was a very brave, very well-known, or very foolish scientist who would travel to either country without benefiting in some way from these roofs. Even for scientists like Kip Thorne, whose interaction took place exclusively outside the established channels, the visa conventions of the master Lacy-Zaroubin general agreement on exchanges made their visits possible, and safe, at the most basic level.

Into the 1970s and 1980s, the top-heavy superstructure of the détente-era formal intergovernmental agreements for science and technology cooperation tended to be a substitute for actual collaborative research, or crowded it out, between working scientists on both sides. High-level meetings often seemed to have been the main product of these programs. Moreover, the very fact that these cooperative programs were housed in these formal, governmental structures made them especially vulnerable to the vagaries of external factors—such as the fragile state of relations between the two superpowers. At least on the American side, their very existence as mutually agreed entities made them objects that could be turned on and off at will by the government. This was certainly a hindrance to the participants but was not always fatal because those scientists who did not depend on the formal programs for financial support or shelter could and often did find other ways to continue.

The experience of the National Institutes of Health (NIH)—or at least the National Institute for Allergy and Infectious Diseases (NIAID), an important division of the NIH—was illustrative of the limited role that the formal intergovernmental agreements played in their international activities. According to Karl Western,

> The way NIH usually operates is that we participate in the formal agreement, but we look on it as an enabling document. And the scientific exchange in the cooperation that gets done, is done outside the formal channels of the Science and Technology agreement....
>
> So, there are two separate questions. One, "What were we doing with the Soviets through the formal agreement?" And the other issue was, "What were we doing with them through normal scientific channels?" because NIH will follow the science wherever it may be. One very unusual feature [of the NIH] is that it isn't a domestic research agency. From the very start after World War II, when we were given extramural authority across NIH, anybody in the world could apply to the NIH for funding. Or they could partner with an American and the more experienced American could submit a proposal which would have a Chinese or a foreign component.

In addition to NIH-specific procedures, there were other considerations. Here again we come back to the interaction of scientific cooperation and foreign policy. "Another unusual feature about NIH," Western told me, "is that anything NIH does overseas has to be vetted by the State Department.... And it doesn't even have to involve funds if they're substantial."[2]

I found this statement surprising. This was not true, except in exceptional cases, of National Science Foundation (NSF)–funded activities that took place outside of the formal bilateral agreements. Western continued:

> But, in practical terms, although nothing [of scientific substance] was happening in the S&T agreement, or in the public health agreement, there were Soviet scientists who were coming here for research training. There were Soviet and Eastern Bloc scientists who were part of extramural grants; they may not have been the principal investigators of it, but they were participating in the grants and, particularly, in some of the networks that we had developed, it was all going on behind the scenes. But if there was one of these proposals, the State Department will look at it; it goes through the scientific arm of the State Department, and it goes to the ambassador; he's supposed to look it and see if there's anything which would be harmful to you as a foreign policy or political interest in the country.

In fact, according to Western, the most scientifically important international work with the Soviet Union came not through its cooperative activities at all but through its intramural training programs for foreign scientists: "Where the first growth usually comes is in the intramural program. There were Soviet scientists and Eastern Bloc scientists who were here as recently minted PhDs at a formative stage in their career to get outstanding research training. And that preceded normalization of relationships by a number of years."[3]

With the fall of the Soviet Union in the 1990s, vast opportunities opened up for new and exciting forms of cooperation. One of these was in the nuclear nonproliferation area. The situation was urgent: Russian scientists in the closed nuclear weapons complexes were starving. The director of one of them committed suicide in despair. There was concern about brain-drain and loss of nuclear materials. In Sig Hecker's story, we have seen how he was able to get permission to travel to the Russian labs when the US government considered it unthinkable, how he actually signed a research protocol with the Russians in spite of his having been forbidden to do that, and how after the National Security Council threw it in the trash and said it didn't exist, oversaw the "lab-to-lab" program that was that protocol's legacy. Hecker was able to do all this for one important reason only: he was not a government employee. Even though the research at Los Alamos was entirely government funded and subject to strict government secrecy, the lab itself was an entity of the University of California system. Hecker said:

> The reason we were able to be very effective—because in the end, what the bottom line was, we were very effective—was that we had a window of opportunity. That window of opportunity was on our side. I wasn't a government employee. My boss was David Gardner, the president of the University of California. And when I said "Look, I want to go to Russia, [but] I mean the lawyers would almost immediately kill that," he said, "You're going to do what? Sign a contract with the Russians? You go do what's right for the country."
>
> So I did what I thought was right for the country. I pushed going to Russia. It wasn't our government that said, "Look, you guys better get over there to Russia." I said, "We should go over there to Russia." And I had the backing of the University of California. It's also interesting that Al Narath, who was the director of Sandia, didn't go on the first trip because they worked for AT&T Bell Labs and Lockheed Martin and even though those contracts were pretty good—they were not for profit—they still thought that was too high risk. But John Nuckolls and I both worked for the University of California, and we went.
>
> The difference then also was that the government wanted to prevent everything, but scientists want to do things; they want to create things. We

were there to create new science, new technology, new things. So we had a totally different approach to it all. I also said in the book that Reagan coined the term, "Trust but verify." As scientists, we developed the trust; we developed the relationships; we were in each other's homes, both their side and our side; we saw each other for months and months and months and months over these years, my fifty trips altogether. And so we said, "Trust and benefit. Do something with that trust." And so we did.[4]

The Civilian Research and Development Foundation (CRDF) was another example of how a nongovernmental entity could accomplish much in terms of both scientific cooperation and public policy that the government could not accomplish on its own because of its rigidity. Established by law as a nonprofit foundation but initially funded entirely by the government, CRDF developed a wide range of cooperative programs with the FSU without rigid oversight, subject to approval only by its independent board of directors. This allowed CRDF to do innovative things that government would not or could not do, such as conduct joint grant competitions on a large scale, provide substantial direct funding for work and instrumentation in the FSU, receive major support from other sources such as private foundations, and even cover expenses of employees of private US high-tech companies to explore industrial R&D cooperation with the region. At the same time, of course, it was subject to an annual funding process, but it was also able to work informally with congressional committees in ways that government agencies found difficult.

Ambassador Eileen Malloy shared her thoughts on another aspect of government inflexibility: obstacles to the flow of information within the government itself. The big picture of US international science cooperation, with the Soviet Union as with many other countries, is one of many cooks and many recipes and little communication among them. One of the largest players is the Department of Defense (DOD), which is often not generous about sharing information with other agencies. As stated by Malloy:

> There's an awful lot of science funded by DOD that is not known to other agencies. I spent ten years working for the State Department's inspector general and a challenge [we faced was that], as we went to each country, we would try to study the totality of US government funds being spent there. It was virtually impossible, because there is no single entity in the US government—not at USAID, not at State—that can provide you all of this information. It comes down to the issue of the ambassador's authority from the president. How can he be on top of it if he or she doesn't even know what's going on? For an example, I was in Australia during the invasion of Iraq. We had a terrible time keeping relations with the average Australian on an

even keel because they were so opposed to the military action in Iraq. So I was always looking for ways to show that there was more to the United States than this one issue.[5]

Using science as an instrument of foreign policy, then, is not as straightforward as it might seem. First, you have to know what's going on across the board—and that can be a big problem.

Lack of Understanding

When I use the phrase "lack of understanding" here, it is not cultural differences or lack of information that I have in mind as much as a willful ignorance or carelessness by people who ought to know better. This malady is a hazard not only for government employees, of whom one might cynically expect such shortcomings, but of scientists as well. In international cooperation, it manifests itself in the scientific community in terms of condescending attitudes toward science, or the possibilities of doing good science, in other countries.

The distinguished environmental microbiologist Rita Colwell of the University of Maryland, a former director of the National Science Foundation, shared with me her experience as a young female scientist who encountered pushback from her professors when she proposed to go to international meetings because, they told her, it would be a waste of her time. And it was not, she told me, only a question of time or gender. In the United States, at least, there was a condescending attitude about science in other countries that was widely shared in the scientific community. She would also get the question "What about these people? [Their] science wasn't very good" from those who wished to discourage her from traveling abroad for science. "But it was very good," she added.[6]

Eileen Malloy gave me a number of other examples from her long experience as a Foreign Service officer. Two of them happen to have involved program managers from the Department of Energy (DOE). Her first story was about the scene around a negotiating table in Moscow:

> I think we did good stuff but what disturbed me was that it was all being run either by DOD or DOE . . . by people, none of whom have had overseas experience for the most part, except in this particular program. They were making all their decisions and their strategies based on this very limited interaction and not underpinned by any kind of regional expertise. Most of them didn't speak the language; they relied on interpreters, and the interpreters on both sides left a lot to be desired.

> I was in one Russian Ministry of Nuclear Energy meeting with a very difficult [Russian] senior official who was the one who would always lecture and scream at us. You know, everything but the shoe on the table. And all the DOE people would get all upset. Once he just walked out of the meeting, and the DOE staff members were all saying, "Oh my God, what do we do? We should give him what he wants."
> And I said, "No."
> They said, "Well, what are you going to do?"
> "I'm just going to sit here because he'll be back," I answered.
> And he did come back into the meeting room and resumed the discussions. And they said, "Hmm . . ."
> When he returned, he had an interpreter who was always used for these bilateral discussions and who was so used to his spiel that in a different meeting, she actually got two sentences ahead of him. I could tell. I was sitting there with this big grin on my face, and all of the sudden this Russian official realized, half listening to her, that she was ahead of him. But the DOE people couldn't do that. . . . So you know, all of this was new, it was exciting. DOE was getting buckets of [US government] money that they never had access to before. They had the technical expertise but it wasn't the whole-of-government effort it really needed to be.[7]

Not only was it not a "whole-of-government effort," but it was not uncommon that many people, scientists and government and others alike, failed to appreciate that they were functioning in a very different culture with different norms and rules. What works in a discussion over a government contract in Washington, DC, as a rule often fares rather badly at a negotiating table in Moscow.

Another troubling—or amusing, depending on how you look at it—story told to me by Malloy drives that point home:

> I had another conflict with the good folks at NNSA [National Nuclear Security Administration]. . . . When I first arrived at DOE, they briefed me on their programs and mentioned that they were providing warm winter uniforms so that the Russian security guards at nuclear facilities would actually get outdoors and do their perimeter checks and visual inspections.
> "Good in theory," I responded, "but do you know the right questions to ask the Russians about procurement of the uniforms?"
> "What do you mean?" was their response.
> I asked, "Where were the Russians procuring these uniforms? Are you providing them?"
> "Oh no," they said, "we give them a sum of money, and the Russians have them made locally."
> "Well, where do the Russians have these uniforms made?" I asked. They didn't believe me when I pointed out that the tradition was that the uniforms come out of the concentration camps, the gulag.

And so then they very happily came back to me about a month later, because there's actually US legislation that requires that we not do this—obtain goods manufactured in the gulag. Well that was not even being thought about as part of the DOE procurement process. . . .

So they came back and they were very happy and they told me, "Well, we asked the Russians about this and they assured us it's not happening."

And presumably that was the truth—or at least they thought so.[8]

Lack of understanding of the other country was not, of course, a weakness specific to the Americans. If anything, there was a much larger vacuum of comprehension on the Soviet side of conditions in the United States. And of course this was a consequence not only of ignorance or cultural boundaries but also of government propaganda—but the latter is the subject of many other books from the Cold War period familiar to many in my field.[9]

Speaking of a lack of understanding, an anecdote shared with me by Tengiz Tzertzvadze illustrates what awaited Soviet exchange scientists upon returning to their home country:

> One interesting event happened in 1984 during the scientific tour. . . . When we visited New York harbor we saw a big picture of President Reagan. . . . Anyone could take a picture next to Reagan's picture and [it] looked like you were speaking with him. The White House was in the background so it seemed that you were speaking with President Reagan in front of the White House. All the members of our group took pictures next to President Reagan's picture. When we returned home, the chairman of the party committee at our institute called me and asked [me] not to show these pictures to anyone. He said that he was called to KGB and asked about these pictures.[10]

I asked him if they understood that the pictures were of them standing in front of a cardboard dummy. "No, they didn't," he answered. "They thought that President Reagan met with us at the White House. I was told, 'How could you, a member of the Communist Party, take a picture with the president of the country that is the enemy of our country?' They wanted to expel me from the party and fire me from my job. I started explaining that this photo was not real and it was [a mocked-up photo]. They would not believe me. This funny story happened to us."[11]

Funding

By including "funding" in this litany of problems, I do not intend to imply that funding was inadequate. In fact, many of the programs—particularly the nonproliferation programs—were very generously funded in gross terms. However, in these and other post-1990 programs—and indeed, from the very beginning—it is my belief that the *structure* of funding for international scientific cooperation in the US government introduced more or less serious distortions throughout the entire system.

Whenever there are special "pots" of money for purposes other than that of a government agency's core mission, the purpose for which that money is to be used is almost de facto separated from that mission. Of course, there are good reasons to create such special funding opportunities, but by and large, when they stick around for too long, they become identified as supplementary to an agency's purpose rather than an integral part of it. From a management standpoint, it might be thought that if that sticky pot's purpose is of such lasting benefit to the agency, such purpose should become a more integral part of the agency's purpose. Instead, the longer they remain separate, the more people in the agency are likely to see its purpose as something apart from the agency's mission.

I will illustrate this abstraction by describing how it worked in the NSF, my "home" agency. Early in the NSF's history, funding for international scientific cooperation was kept in a separate category, perhaps because at the time, the agency's mission was more narrowly thought of as promoting the health of science in the United States than it is today, when all seem to agree that "science knows no borders." Whatever the original reason, that separate "pot" of money for international activities at NSF grew and became physically identified with a separate NSF division, the Division of International Programs, or INT. INT's ambitious leadership assiduously increased its funding from year to year, as all diligent division directors do, until other NSF divisions, particularly those funding discipline-oriented research, began to take note that there was somehow an international "empire" at NSF that was being run independently of, and often in competition with, the NSF's core research programs—which was indeed the case.

Even worse, because the cooperative research projects funded by INT were evaluated separately from those in the research programs themselves, there was a widespread perception in the latter that the international projects were not of the same quality or relevance to the disciplines as their own. This probably had some merit, though that did not mean that the international projects funded through INT were not meritorious—they were just different, separate, not seen as part of the NSF's core mission, and perhaps as less scientifically competitive than those funded through the research programs themselves. Ultimately, these attitudes became so entrenched and intractable that the international division was drastically scaled back to become part of the NSF director's office—where such staffs are usually located in government agencies—and its funds restructured and reprogrammed to be more in line, as it was thought, with the agency's mission as a whole.

I tell this unfortunate story here not to criticize or complain but to make a point: The structure and nature of funding for international scientific cooperation strongly influences its use and the attitudes of the people who use it. In the broader context of international scientific cooperation government-wide, a similar situation came to develop over the course of the sixty-year story of US-FSU science cooperation narrated here. At first, in the Cold War years, the primary vehicles for US-Soviet cooperation were the exchange programs, which were funded separately; the interacademy program, for example, was funded by the NSF Division of International Programs with funds that were more or less specially reserved for that purpose, originally at the request of the Department of State. With the advent of the large, showy formal intergovernmental agreements of the détente era, there was a need within the agencies to create specially sheltered funds—particularly for management and travel—to operate their cumbersome infrastructures. Meetings of various kinds and delegation visits (both management and scientific) were much more common in these programs than before, and agency funding patterns began to reflect this new reality. By the time I arrived at the NSF in 1979, the annual budget for the US-Soviet program consisted of a special appropriation of more than $3 million, as compared with the approximately $10 million budget for the entire international division—and this did not include staff salaries and travel costs, which were in yet a separate tranche. Our division director was not amused, I am sure.

Other agencies were not nearly as fortunate. As a rule, these agencies had no specific funds reserved for US-Soviet cooperation, so when a joint commission or joint committee agreed that such-and-such cooperative research should take place, the agencies complained bitterly that they had been given an "unfunded mandate" that required of them something that was not possible. Unless an agency had discovered some truly unique phenomenon or research in the FSU that was vitally important to its core mission, there was simply no money to do anything else. Examples of the former would be, for example, the expeditions of the US Geological Survey to Lake Baikal in Siberia or some of the work of the National Institutes of Health on AIDS and tuberculosis in Russia. But these cases were more the exception than the rule; across the broad front of the eleven intergovernmental agreements of the 1970s and 1980s, similar compelling incentives were not that common.

In the 1990s, these structural anomalies became much more substantial, with more serious results. On the one hand, government funding in many civilian agencies for US-FSU cooperation shriveled; in 1991, my budget at NSF for the new US-Russia Basic Sciences program was about $135,000—in NSF terms, virtually zero. On the other hand, massive new programs came on line, of two types: Soros's $100 million International Science Foundation (ISF), which was targeted toward assistance for civilian scientists in the FSU; and the similarly massive international and national nonproliferation programs, focusing specifically on "former" weapons of mass destruction (WMD) scientists. With this volume of third-party funding directed in general at science in the FSU, and with its science infrastructure falling apart, it was understandable why many agencies relegated cooperation with that region as integral parts of their mission to secondhand status at best.

In the area of nonproliferation in particular, new special-purpose programs sprang up to address specific concerns. Several were directed at the problem of biological weapons in the FSU. Thus, for example, there was the Bio-Industry Program of the Department of State, the Biotechnology Engagement Program (BTEP) of the State Department and the Department of Health and Human Services (HHS), and various programs under the auspices of the Department of Defense. Where these programs generated special "pots" of money for mission agencies, as did the BTEP program for HHS, there were both new opportunities as well as new dependencies on special outside funding created. The International Science and Technology Center (ISTC) and the Science and Technology Center Ukraine (STCU) offered additional opportunities for civilian agencies and others to engage scientifically with scientists in the FSU, provided they could show that the latter had been involved in WMD research. These activities were very important in terms of global security, but they also no doubt represented that much of a distraction or redirection of the host agency's scientists from other possible areas of engagement.

Thus, it became so customary to rely on these special tranches of money for scientific cooperation with Russia or Ukraine, for example, that when these funds themselves began to dry up in the second decade of the 2000s, the agencies saw little incentive to undertake new initiatives with these countries. If there was no separate funding, there was simply no funding at all. To be sure, the antagonistic political climate of those times did not help, either.

F. Gray Handley, a senior international manager at the NIAID at the National Institutes of Health, reflected on these problems in our discussion:

> The problem with those agreements—and all of these kinds of agreements—is that they get signed at a political level. As you know, they have no funding associated [with] them. So, unless there is a personal relationship and a driving scientific need and something very special that you really want to access, like a population or something of that nature, it is very hard for an institute to find the resources, because we [distribute] almost all of our money based on peer review of applications. There is almost no money that a director has available to her or him to put into a targeted bilateral program. That's just very, very hard to do because of the way we manage our resources, which is good. I mean, that protects them from being misused, right? . . . But it means that when we're handed a mandate like, "Do this artificial heart thing," it's hard for us to find much money even to have meetings. In the old days, it was a little easier than it is now; now it's even harder.
>
> The only time that I've really interacted very much with the former Soviet Union is when it became Russia, and what happened then, of course, is that we, like all the other technical agencies, were asked by the State Department to find ways to work with the former Soviet Union and with Russia. . . . But for example, when BTEP money came around . . . we did look for opportunities and that provided us a tool that we didn't otherwise have, so we were able to look for opportunities. In addition, there was money from the ISTC and the STCU. These mechanisms, when they were set up, opened up an opportunity for NIH that we wouldn't otherwise have had. . . . And since these resources [were] available, [that] made us take a second look at what the opportunities might be. . . .
>
> Also, we were asked to help with the effort to move the bioterrorism folks in Russia who had been working on microbes and infectious diseases into other areas of health science, and that is, of course, what we do. So for those ten years or so, we had access to resources through ISTC and through SCTU and, to a lesser degree, through the BTEP and other State Department programs, and we could encourage scientists to find partners in Russia and the former Soviet Union. And many did. . . . Those organizations would put out calls, we would review applications, we would score them, and we would track them and manage them scientifically, and in some cases, there were resources available for our intramural scientists, which was really a boon for many of them. And so, I think some good work got done. I'm not sure it was very extensive, but it wasn't very much money either, relative to our institute's budget. So I think it was a useful program, but only as long as those [additional] resources were available.[12]

Logistics

Logistics are always a major issue for any scientific expedition in the geosciences, where the main focus is on researching a particular location. Usually, those locations are in far-off places. Some of the largest, most ambitious cooperative projects with which I had any direct contact were of this kind. Almost all of them were on the territory of the FSU—in northern Siberia to examine permafrost, in Kyrgyzstan to monitor earthquakes and other seismic phenomena,[13] and at Lake Baykal to research the Earth's early climate as well as plate tectonics.

During the Soviet period, if such a project were approved—which was always very complicated—many layers of review were required. While the requirements and delays could be immensely frustrating, the Soviet institutes were very capable in doing this kind of research and the expeditions usually went off reasonably well. After 1991, however, as we have already heard, "all hell broke loose." Institutes lost funding and resources, the lines of authority to issue permits were unclear, transport and other services became "privatized," and especially in the more remote areas perhaps, *bakshish* was *de rigueur*.[14]

During those times, it was only the most hardy (or foolish) American scientists who attempted to go it on their own. One of the hardy (and, I would add, lucky) ones was Julie Brigham-Grette, whose story about the scientific outcomes of her expedition to Lake El'gygytgyn in remote northeastern Siberia was detailed in chapter 7. The nonscientific side of her work there was, to me at least, as entertaining and in some cases appalling as her scientific results were impressive and included outright bribery, extortion (failed, but at the cost of being stranded in one of the most remote places on the globe), and even the "murder" of a key scientific instrument. Her story, "The Expedition from Hell," appears in chapter 11.

Idiosyncracies of FSU Science

Although relatively few American scientists mentioned the quality of Soviet or Russian science as a negative factor affecting success of their cooperative efforts, there were exceptions. A notable one was in the health sciences. Perhaps, in part, this area of perceived weakness was due to the terribly destructive legacy of Lysenkoism on basic research in Soviet genetic science. But there were other factors in play, some of them structural. Karl Western of the NIAID told me that with Russia per se, we have not really had what I would regard as a major success.... Many of their health scientists are not well-trained.... Many of them are not interested in research, many of them have their old way of doing things and they're not interested in changing. They don't have a peer review system, and the bureaucracy or government is so corrupt, they take 50 percent off the top before they accept the award and, you know, unlike the NIH which, if the award is to Hopkins, you know, it stays in Hopkins.... So, I think what it comes down to is, it's easier to do the same experiment somewhere else, than go through all the time and trouble and frustration of trying to do it in Russia."[15]

When I asked him whether there was some kind of comparative advantage in Russian health science that made it of special interest to the United States, Western commented further:

I'm going to take it from the viewpoint of our institute, I would say the answer was no. The impediments were considerable. I've already said that they don't train their physicians or their caregivers, whether they're doctors or nurses, or veterinarians, in the science, they train them to be good ... practitioners.... So the NIH would occasionally be attracted to the absolutely brilliant individual scientist who had original ideas and thoughts, but we had a lot of trouble dealing with the biomedical and public health people because they were not up to standards. Furthermore, the Russian economy had been running on ether, for a very long period of time, and the biological scientists were not the most favored field ... so the Soviet scientists and their allies, as we went into the '90s, were working in 1970s facilities....

They were not up to standard, technically, and by and large they weren't safe working on the infectious pathogens that we were, like HIV or TB, particularly, and DRTB, so the dilemma was that our modus operandi is to try and do as much of the research in-country as we can. It's a lot easier if there are laboratories, scientists that are well-trained in laboratories that work....

So I think it's the problem of biological scientists are not at par, they're not trained in research, they're not interested in research. After the breakup of the Soviet Union, there was only one newly independent state that reorganized its science system and integrated research into their new health system; that was Republic of Georgia.[16]

Western went on to clarify that it was not necessarily only the equipment or the quality of training that was an obstacle but also the structure of Soviet science itself: its rigid vertical division into disciplines, which was a function of the vertical organization of Soviet science into enormous, highly specialized institutes. By contrast, NIH took a more holistic, cross-disciplinary, problem-oriented approach, focusing not on disciplines but on problems—in its case, diseases:

That was another feature of Soviet science; they did have these disciplines and they would want to collaborate in the discipline, and we would say, "What disease or what pathogen are you going to apply the discipline? Because we don't have anybody who specializes in that discipline . . . " So there's not a lot of common ground. . . .

You know, we're working on Ebola. And we're working on Zika. But we apply all the disciplines to the one disease or pathogen. That's of importance. Now, what we find out may be applicable across other flaviviruses and to other diseases, but it's really a multifaceted targeted effort, [whereas] in the Soviet system, if you work outside the discipline of the institute in which you're located, you're suspect.[17]

Human Rights, Anti-Semitism, and Ethnic Discrimination

During the Soviet period, especially the 1970s and 1980s, scientific cooperation between the two countries was beset with all manner of problems relating to the Soviet Union's repression of dissidents and persecution of minorities, in particular Jews. Some of my interviews touched on these subjects.

The persecution of Soviet nuclear physicist and human rights activist Andrey Sakharov was the most dramatic and well-known example of Soviet repression within the scientific community. During the turbulent period of the 1970s and 1980s, American and other scientists were seized with doubts and arguments about whether it was morally or otherwise appropriate to participate in scientific cooperation with that country at all. The Soviet government's persecution of Jewish scientists, particularly with reference to their desire to immigrate to Israel or elsewhere, were also an enormous cause célèbre of concern in the US scientific community. Many advocated a total boycott of all interactions and travel; others argued that only through continued contact and speaking out could foreign scientists make any impact on the situation.

Kip Thorne had been traveling back and forth to the Soviet Union for many years by the time these issues erupted. His purpose was a singular pursuit of advanced scientific research in astrophysics, an area in which Soviet scientists such as Zel'dovich and others were undisputed leaders. Many were Jewish.

Thorne found himself in the middle of controversy among American scientists, but he persevered. I asked him, "Did things keep going on?" during this period. He replied:

> Yes, as far as I was concerned. . . . This is when we were getting LIGO off the ground and this was absolutely crucial in that era as well. . . . We continued. I remember things were a little more difficult, but not a lot. For me, there were people telling me that you shouldn't be going to Russia anymore. . . . And

there were others who were going there to support the refuseniks. So within the community there was a lot of turmoil over "Do you go or don't you go; how do you support the dissidents?" ... Especially in the physics community. And Sakharov was in exile. [Vitaliy] Ginzburg was his principal supporter. Sakharov was a close friend of mine. And Ginzburg was a very close friend of mine. And so this was all pretty personal. [Leonid] Ozernoy, who had been a collaborator of Ginzburg's, was a refusenik, and he had been thrown out of the institute. I had discussions with Ginzburg. He said, "Look, Ozernoy will get an exit permit within the next two years. Don't worry about him; he's the lucky one. The rest of us are trapped here."

There was a lot of bitterness; there was real turmoil. My own collaborations moved forward. I walked a bit of a tight rope then over the refusenik business. And Sakharov. And I carried personal messages back and forth for Sakharov. But my own collaborations continued vigorously."[18]

Carrying messages back and forth for Sakharov was no small thing, and while it was not known at the time, it is a powerful reminder of the importance of keeping channels of communication open in the scientific community, even in the worst of times.

Glenn Schweitzer was deeply involved toward the end of the Sakharov crisis, and indeed in bringing it about. By the mid-1980s, all contacts between the US National Academy of Sciences (NAS) and the USSR Academy had broken down over the Sakharov affair. Schweitzer recalled:

In 1985, I accepted a position at the National Academy of Sciences with responsibility for directing the academy's exchange programs with counterpart academies in the USSR and countries of eastern Europe. My first assignment was to put back on track the interacademy program with the Russian [Soviet] Academy of Sciences that had been derailed several years earlier due to Soviet mistreatment of Andrei Sakharov and of Jewish refuseniks, who were losing their jobs due to their complaints over discriminatory practices by the directors of their institutes and were being denied permission to leave the Soviet Union.

I promptly left for Moscow, with instructions to gain in writing a commitment from the [Soviet] Academy of Sciences to address such human rights issues. With such a commitment, we would be prepared to renew the exchange program based on the language included in previous interacademy agreements. After three days of discussions with Soviet officials working within their academy, we decided to have a farewell luncheon.

During the final discussions at the luncheon we agreed to retain most of the text of the previous agreement, but the Russians pressed hard to insert previously proposed language, such as "will comply with the regulations of each country" and "will not interfere in the internal affairs of the host country." The previous day, I had consulted with Washington, and I adamantly rejected these insertions at the luncheon since they could provide reasons for Soviet crackdowns on contacts with foreign scientists. At the luncheon, it became clear that the Soviet Academy was eager to resume the interacademy exchanges and they were prepared to withdraw their proposals for insertion of additional language. However, I still had an unfinished piece of business.

The Soviet official who led the Soviet team was a well-known senior official of the Soviet security services, and I assumed he had authority to make a commitment if we could find acceptable language to include in the document. He thought we had closed the deal, and he was surprised when I returned to an earlier request, indeed a demand, that we have an insertion recognizing the importance of human rights in the agreement. After thirty minutes of back and forth, we came up with a sentence that stated that the two academies would "consider the environment for carrying out exchanges" in implementing the agreement. I scribbled this sentence down on a paper napkin and handed it to him. He thanked me, we shook hands, and I returned to the hotel and prepared for my departure for home.

Several weeks later in Washington I received a message from Moscow accepting the agreed text, including the language written on the napkin. Our exchange program was soon back on the rails and has continued in various forms ever since.[19]

As it turned out, this new interacademy agreement was a harbinger of Mikhail Gorbachev's historic but failed effort to reform the Soviet system from within, perestroika. As the regime's liberalization under Gorbachev proceeded, the walls holding back the flow of Jewish scientists abroad also began to fall, leading ultimately both to the fall of the Berlin Wall, presaging the end of the Soviet empire itself. These events also led to the regime's release, also approved at the highest level in his retelling, of Boris Shklovskii to the University of Minnesota, where he established a haven for other distinguished Russian physicists, which came to be known as "Moscow on the Mississippi." Shklovskii was and remains passionately opposed to boycotts, be they in support of Jewish scientists, dissidents, or, more recently, in protest of Russia's aggression in Ukraine in 2014. He explained his position to me in the following words:

> Science has no borders. Enforcing a new border is counterproductive for any side of science, Western or Eastern, any side. And so if you say, "I will enforce no cooperation between the East Coast and the West Coast," it would be a disaster. So it's not about Russia or the United States; it's not about two different systems. This is about science. And for science, whenever you divide it, you inflict damage on science. Of course, this also has a political dimension. For example, the fact that I hiked with this guy at Lake Sevan in 1979[20] maybe changed my understanding of the West. And so of course it had political dimensions. I don't know about now; I don't think so. People do travel. That's not that important. It's completely different era.

> Any cooperation in science is natural and productive. Any blockage of science is unnatural and unproductive, and . . . I'm against any attempt to block Russian science now. And I'm very strong on this point and I would like to state it for the record.[21]

Of the standoff in US-Russia relations in the wake of the 2014 Russian aggression in Ukraine, just prior to my interview with Shklovskii, he added:

> I am very, very sorry that the American government and American authorities are trying to block the existing, very well established, rules of cooperation. I can give you crazy examples . . . for example, what happened at Fermilab in Chicago, which has enormous computational competence, because taking data from accelerators requires enormous computational effort. And it happened that such expertise exists in Novosibirsk, at the Budker Institute of Nuclear Physics. And lots of Russians came and worked in the core of the computational part right here. (It's not my field, but that's what I hear from friends.) . . . And again, when your technology is not good, you try to compensate with your mind, and when your computers are not good, you develop tools, programming tools. And so they actually created this kind of competence at Fermilab, but then people start blocking guys from going back and forth rather than trying to keep this working, and then things fall apart.
>
> That is what happened about a year ago. I'm not involved in this myself, but whenever somebody stops here from Russia, from Novosibirsk, and tells me this story, I am very, very sorry that such things are happening here. I don't think it's productive. I don't think that pressing Russia that much generally speaking is productive. And I do think that is dangerous too. Because . . . this is political. So people in the foreign office [the State Department] think that they're helping to get rid of Putin or something, but if they get rid of Putin they can get somebody much worse. . . . Everybody thinks about today's political statement and how beautiful he will look when he gives the speech, and nobody thinks ahead.[22]

In the interviews, I heard a great deal about anti-Semitism from the émigré Russian scientists I interviewed. That topic is probably worthy of a separate book. In chapter 5,[23] we heard Shklovsii's story about how Jewish scientists were limited or prohibited from attending some early US-Soviet physics meetings taking place in the USSR. Maxim Frank-Kamenetskii, now at Boston University, told me the following story about how the famous Soviet biologist Aleksandr Bayev personally blocked him from visiting an institute in France:

> There was a guy who is not alive any longer, his name was Claude Halen. He was [a] very, very respectable French scientist. . . . He was head of some small institute in Paris. We were very close colleagues. . . . It must have been in 1978 or 1980. . . . He sent an official letter inviting me, because, he said, "I want to

establish collaboration between French science and Soviet science and I want you to come, to visit us here and we will discuss this, how we will collaborate." And he sent this paper to me, or maybe even to [the] institute director, somehow on the institute level.

My director at that time was a man named Makulski; he was really excited about this. . . . Our institute [of molecular genetics] had just been moved from the Ministry of Atomic Energy to the academy.[24] And we got this letter. And he said, "OK, I fully support this; I will send [a] letter to communicate to the academy administration to arrange your trip." And after a while, he invited me to [see] him again, and he said, "You know what?" He showed me the letter that [he] had written to [Alexander] Bayev, academician Bayev, who was academician-secretary of the academy's Biology Division. In this letter Makulski said, "I fully support this, I want you to support Maxim that he goes to [Paris]." And he showed me Bayev's resolution, which was already a matter of principle: "I support this collaboration but please find another person to participate from our side."

"Because you were Jewish?" I asked Frank-Kamenetskii.

"Yes. . . . It was pretty traumatic. Because it was not, like, from the KGB. I knew Bayev personally, and he knew me."

"Do you think it was Bayev's decision?" I asked. "A hundred percent," Frank-Kamenetskii answered. He continued:

> Why on the earth did he have to take this on himself? Otherwise, he might call to Makulski and say, "You know what? It will not work." But he did it [this way]! He did it explicitly. And this guy, my director, he was put in such a position, he wanted to show me that it was not him, that he did not play this . . . game with me, because he had defended me already. I was a very, very prominent figure in the institute and in the scientific community. He wanted to be a member of the academy, and he knew that I could influence this, I could support him or not; I am not like, you know, nobody. . . . I had very good connections with many people; like I knew Gel'fand personally very well, and Zel'dovich and many others. And so he did not want me to sink [his potential membership]. He played a double game, that he supported me, and he showed me this because of that, because he wanted to not be blamed. But it was really, really shocking, absolutely shocking, because . . . Bayev was just an anti-Semite. Simple. That simple.[25]

This type of incident was not at all unique and not at all uncommon. The physicist Lev Okun', who was also Jewish and who also, like Frank-Kamenetskii, had a reputation for being independent-minded and outspoken, was also denied permission to travel abroad in 1979 to a scientific meeting organized under the intergovernmental Science and Technology agreement, causing a major scandal and what turned out to be a permanent suspension of its activities by the American side.[26] Both Okun' and the

Americans had expected him to attend the meeting, and I can imagine that there must have been a similar scene of an unexpected higher-level veto in his institute director's office as well. Such last-minute reversals may not have been very common, in part because few actually attempted to challenge the system in this manner, but the principle—that Jewish scientists do not travel abroad or attend international meetings even at home—was very widespread. In a sense, the entire refusenik movement was the general case of this discriminatory principle writ large: if you want to leave the country, don't even think about it.

Anatoliy Logunov, the longtime rector of Moscow State University (MGU), had earned a terrible reputation for his role in blocking the immigration to Israel of many Jewish members of the MGU faculty. According to Kip Thorne, however, the situation was more nuanced than most observers might have thought. Said Thorne:

> When I would go to Moscow, I would generally meet with the rector, who was Anatoliy Logunov. His research was in my field, in general relativity, but he had developed his own theory of gravity. And just because it was him, I tried to understand it, and I understood that it was just general relativity but with some physical reinterpretation. So on one occasion in the late '80s—and it may have been the same occasion when [Vladimir] Braginsky had organized an honorary doctorate for me from Moscow State University (the person who got it immediately before me was Angela Davis)—Braginsky organized the usual audience in Logunov's office, a private visit, and Logunov sat down to explain his theory of gravity to me. And so I told him that it was really just an equivalent to general relativity with just some reinterpretation, and then we got into some argument over it and Braginsky was translating. And the sweat started to roll down Braginsky's face. I was telling Logunov . . . at one point I had to say to Logunov, "If you continue to insist that this is an entirely new theory, you'll make a fool of yourself with the physics community in the West." I mean it was a very frank discussion and Braginsky was just sweating.
>
> Afterwards, we left and Braginsky was shaking, and what was Logunov's response? He arranged for me to leave the Soviet Union as a VIP at the airport. I didn't have to go through the usual procedures. He treated me with the utmost respect and there were never any repercussions for Braginsky. Logunov, I think it turned out, actually appreciated having a frank scientific conversation. And so I gained increased respect for Logunov. I didn't have a lot of respect for his science, but I respected him as a man and as an administrator. He continued his vigorous support for research in this field for Braginsky and so forth. He was regarded, I think by people in the West and by some of the physicists in Moscow, in a very negative light. . . . So it was interesting to me to see this man who was vilified. It was because of the refuseniks. He was making decisions that people didn't like and was not letting some people, refuseniks,

go. But he seemed to do it in as humane a way as he could, he seemed to me to be functioning under the circumstances in a more or less reasonably admirable way—to the extent that I knew what was going on.[27]

Those are certainly the most positive words I have ever heard about Logunov, though even Thorne's story is a very complex mix of impressions. Still, Logunov was savvy enough not only to govern a university for many years but also to recognize good science when he saw it, to enable it as best he could, and even silently to acknowledge that his own proudly held scientific theories were not all that he made them out to be. None of this is an excuse for enforcing government policy of blocking Jews from emigrating or even for personally held views about Jews, but it is an important part of the broader picture.

While discrimination against Jews was often crude and widespread throughout the FSU, both in terms of access to international science and in other ways, they were not the only ethnically (or religiously) disadvantaged group of scientists in the Soviet Union. In the exchange programs and cooperative research programs of the 1970s and 1980s in which I had direct experience, it was the uncommon scientist from Armenia, Georgia, Central Asia, or even Ukraine who was able to participate. The major exception to this rule, in my own experience, were the metallurgists from Kyiv, where the work of the E. O. Paton Institute of Electric Welding was second to none in the world and of much military and commercial interest besides. No scientists whatsoever, as far as I can recall, ever participated in the exchange programs from the Baltic republics of Lithuania, Latvia, and Estonia. When I did come across the odd scientist with an Armenian, Georgian, or even Assyrian surname, for example, it was most likely that he (never she) was working at a Soviet Academy in Moscow.

Yaroslav Yatskiv shared with me a fascinating story how he became a "political" scientist, not in the conventional Western sense of the term, but in the literal sense. His ethnic Ukrainian background and family history had a lot to do with it, and I cite it here because it also tells an important story:

> My grandfather and grandmother loved Maria Theresa of the Austro-Hungarian Empire; they respected her very much, but they really hated the Polish Empire because the Poles were very prejudiced towards the Ukrainian population. After that, some of my relatives welcomed the Red Army with red flags on September 1, 1939, and half a year later, they were sent to Siberia like everyone who did not fit in the political ideology of the Soviet Union.

> This is my political background. For instance, I was not admitted to the Radio-Technical Faculty of the Lviv Polytechnical Institute because there was censorship not to admit "unreliable" people. When I graduated with honors from the institute, I had a Lenin scholarship, and the best graduates were selected to join the space military program in Moscow, at the V. V. Kuybyshev Military Engineering Academy. This was the beginning of the space race between our countries.... I was not selected for the military space program, and I was very upset about that. Twenty years later, however, I was so glad that I had not been selected. All my friends that went there are not alive any more. They were sent to the so-called Scientific Measuring Points in Central Asia, without water, without food. The work there was extremely difficult, without training. I always say that everything they [the Soviets] did against me was hard here, but after some time, it turned to my benefit.[28]

Intelligence Gathering and Secrecy

In a sense, this topic is unremarkable, since it is widely thought, and probably true, that at least at one time—prior to 1991, at any rate—all Soviet scientists traveling to the United States or to other nonsocialist countries were given briefings and, in many cases, tasked by Soviet intelligence authorities before the visits, and then required to submit reports afterward. Like Communist Party membership, this was essentially a requirement. Whether all Soviet scientists fully complied with this policy's intent is another matter, and I personally doubt that this is the case, as scientists everywhere are a diverse lot and not necessarily inclined to do everything that authorities require or expect of them. The US government, though its tactics were less heavy-handed, was also eager to glean information about the state of Soviet science and technology in many fields, including military applications, from the experience of American scientists with their Soviet counterparts. This is no surprise to anyone. But the direct testimony I heard through my interviews put meat on the bones of these perceptions that only they could provide.

Moisiei Kaganov, a senior physicist of the Landau School—who never visited the United States until his emigration many years later—told me about how, in the early 1960s, he was put in charge of organizing an international conference on solid state theory by the famous physicist Ilya Lifshitz. (Another story told by Kaganov about this conference appears in chapter 11.) In the following excerpt, Kaganov recalls his discussions with "Mikhail Mikhailovich," who was the deputy director of the Khar'kov Institute of Physics and Technology for "discipline":

We called him "Mikh-Mikh" behind his back.... The venue of the conference was Moscow.... The conference was already planned. I had to go to Moscow to solve any last-minute issues and stay for the duration of the conference. Mikhail Mikhailovich invited me to his office before I received my travel documents. Mikh-Mikh pulled out a piece of paper with a mysterious look on his face. He showed me a very long formula and started babbling something about asking Western scientists about a certain compound. I did not understand it, nor did I want to understand what I had to find out. I got angry and asked him to stop childish spy games. He did not insist.[29]

To be sure, not all Soviet scientists at that early date, or later ones, shared the same sense of bold irreverence as Kaganov. But the fact that "Mikh-Mikh" did not pursue the matter itself suggests that while the tasking itself was required, compliance was not always a given.

In the United States, the gathering of intelligence from American participants in the exchange and cooperative research programs was a highly political and emotional issue for many scientists. My first exposure to it was in my work at the NAS in the early 1970s, when many university departments were categorically rejecting any defense- or intelligence-related research. The very idea that scientists would be briefed and debriefed before a foreign trip anywhere was anathema to many and for understandable reasons. I had very mixed feelings. I had to acknowledge that one of the key rationales for me about US-Soviet scientific cooperation at the time came from my Cold War "know thy enemy" mentality. Moreover, the NAS exchange program, though carried out by a proudly independent scientific institution, had strong roots in the post-*Sputnik* idea that as a nation we needed to understand what was going on in Soviet scientific institutes, as well as in the noble impulse to pursue cooperation and dialogue with esteemed scientific colleagues wherever they may be. But in the passionate atmosphere of resistance to the Vietnam War and suspicion of US government motives, I was deeply conflicted about the whole issue.

Somewhat to my initial surprise, I discovered that the NAS's policies (or at least unspoken practices) made my personal views moot, or at least irrelevant to my work. We required post-trip reports of all US exchangees, and these were regularly shared with the State Department. I am sure that the destination in that agency was its Bureau of Intelligence and Research, which in turn was of course closely linked with the intelligence community. We also received periodic visits by a very curious man named Mayo Stuntz, who, as it was explained to me, was a CIA officer who worked out of a tiny office above the old CVS drugstore on the corner of Nineteenth Street Northwest and Pennsylvania Avenue. Mayo was a very cordial man who wore a fedora that was too small for his head like a character straight out of a bad spy movie. He asked for the itineraries of our visiting Soviet and East European scientists, and we were informed by management that since they were in the public domain, we were to share them. It was perfectly obvious that the purpose of the exercise was to give their US hosts' coordinates to the intelligence community so they could be contacted by local field agents. Presumably this information was also shared with the FBI, since there was a revered high-level interagency committee called COMEX, the Committee on Exchanges, chaired by the intelligence community and whose proceedings were classified, which had to pass on every incoming exchange visit and coordinated policies and actions on perceived threats to national security arising from the exchange programs.

At the NAS (and later, in my work at the NSF), our interaction with COMEX was strictly prohibited. It was the State Department's Soviet Desk that, unseen by us but with our full understanding, handled those transactions. We would submit proposed itineraries of our foreign visitors to the Soviet Desk, and then weeks later, we would get our guidance from the Soviet Desk. Sometimes the visits were approved without comment, sometimes they were forbidden, and on occasion we were informed that a certain visit on a given itinerary was not allowed.[30] That was the required procedure, not only in the scientific exchange programs, but also with regard to scholarly visits conducted by other organizations, such as IREX and the Fulbright programs. And though it left a bad taste in my mouth, it did make a certain amount of sense, for nobody in their right mind would want to have been the cause of a breach of secrecy regarding a sensitive military technology. Now if this sounds like an apologia, perhaps it is. But as it is an integral and essential issue to understanding the cooperative programs covered in this book, I will continue.

There was a natural division of labor within the national security community regarding our foreign guests. The CIA's interest was intelligence collection; the FBI's, in preventing intelligence collection by the other side. At times, I was to learn later, the two agencies clashed vigorously about specific proposed visits. The CIA wanted to know what a Soviet or Hungarian scientist knew, either about their work on defense technology or what they knew about ours, while the FBI's mission was to prevent anyone who was identified as a potential security threat (which was almost routinely any visitor[31]) from getting access to our secrets.

This natural internal structural tension usually worked to the benefit of the important principle of open scientific communication, in the sense that most visits were ultimately approved. We drew from that the important lesson that the US government's interest in collecting sensitive scientific and technical information about its adversary was more compelling than its interest in preventing the unauthorized leak of such information. Later, as a government employee at the NSF, I was to understand better—though still at arm's length—that in many respects, the intelligence community, including the CIA and DIA (Defense Intelligence Agency), were often in effect our "allies" in COMEX and its successor committees in that they objected less often to scientific exchange visits than did the FBI, the Commerce Department (with its responsibility for administering export controls), and the military research agencies (which objected almost as a rule, but not always, if there was something they really wanted to know). As at the time I did have secret security clearance (the lowest level), I cannot comment further here on these issues; the reader will just have to take my word for it, such as it is.

While it was true that many American scientists, either as domestic hosts or travelers to the communist bloc countries, were briefed and debriefed by US government officials, it is also probably true that this was not a universal practice. Based on my experience, this was probably most prevalent in the early exchange programs prior of the Cold War period prior to 1972, discussed in chapter 1, than those of later years, in which the very numbers of American participants skyrocketed by an order of magnitude. It was also surely more prevalent in certain areas of science than others.

In physics, for example, the Soviets and Americans were world leaders in certain areas of theoretical physics and nuclear physics, and participants from both countries were often engaged in defense-related research even while being university professors or Academy of Sciences researchers. The same might have been true in some areas of materials science as well as chemistry (in particular those fields of chemistry related to explosives and rocket fuel). Microbiology, as it potentially relates to germ warfare, was probably another such area; "civilian" Soviet institutes such as the Shemyakin Institute of Bio-Organic Chemistry were well known to have been funded generously by the military. Experts in hydrodynamics, with its applications to naval design, may have been subject to these procedures. On the other hand, geologists, soil and earthquake engineers, petroleum chemists, zoologists, botanists, and anthropologists—to name a few—must have been of lesser interest in this regard. Again, I must acknowledge that these are unsubstantiated statements because I never had—or desired to have—access to such information, though the frequent feedback I received over many years from the scientists themselves gives me reason to think that these observations are correct.

And if and when American scientists were selected by the national security community for information, there was certainly not universal compliance. Gravitational theorist Kip Thorne shared with me the following two curious stories:

> In the late '60s and early to mid-'70s, whenever I would go to Russia, friends of mine at JPL [NASA's Jet Propulsion Laboratory in Pasadena, California] would say that "they"—I don't know if it was the CIA or the FBI—were coming around asking about me. They said, "We told them you would refuse to cooperate if they went to you directly." So they never came to me directly.
>
> It went a little further. John Wheeler, who had been my mentor, was on the team to set the negotiating strategy with the Soviets about nuclear weapons arms control. Wheeler and I were close personal friends. Wheeler had a discussion with [Secretary of State William] Rogers . . . about me and told Rogers that Thorne is a crucial conduit between the physics communities of the two countries and he should just be allowed to do what he does. I don't know whether that ever led to anything or not, but Wheeler just told me about that conversation. Anyway, they never came around to ask me about what I did in Russia.
>
> But on the other hand, the FBI was charged with keeping track of the comings and goings of all Soviet visitors. There was a particular [man] who was my FBI person; my contact, a man named Mr. Bevins from the Los Angeles office. Whenever Braginsky would come, usually just after he arrived, Mr. Bevins would come to see me about Braginsky. And I would say, "Yes, he's arrived; this is his itinerary," and Bevins said that he might come back around the end of Braginsky's visit just to make sure. So he was doing due diligence.
>
> But on one occasion, Mr. Bevins came and he said, "Can I come into your office to talk about Braginsky?" And I said, "Sure."
>
> I knew Braginsky was in the office. So I ushered him into the office, and I said to Braginsky, "This is Mr. Bevins from the FBI; Mr. Bevins, this is Vladimir Braginsky. Why don't you talk to each other directly?"
>
> Dead silence for about thirty seconds. And Mr. Bevins lifted his pants cuff and pointed to his leg and said, "Look, I'm made from flesh and bone just like you are." And then they had a kind of strange conversation. I then challenged Braginsky to do the same thing with the KGB, but he never did.[32]

Surely there is much more to say about the issue of intelligence gathering and scientific cooperation. But both because of lack of direct exposure or my oaths of (extremely low-level) secrecy as a government employee, my testimony on this subject has to end here.[33]

Finally, one of my favorite stories was shared with me by former US Foreign Service officer John Zimmerman. It is about a quarantined blackboard:

> The S&T programs not only supported good science and an exchange of ideas between American and Soviet colleagues but also gave us a better idea of the other side's intentions. Occasionally, the discussions became a bit awkward because of the differing ways in which each side handled sensitive topics. The one which most often comes to my mind deals with how you discuss certain sensitive areas of thermonuclear fusion because you can inadvertently slide into topics related to warhead design. In this instance—and I hope my memory is correct on this—the US classified shapes and designs, while the Soviets classified materials. In any case, during a bilateral meeting in 1977–78 between researchers on this topic, a Soviet scientist drew a shape on the blackboard, which was harmless in his view, but which greatly alarmed the Americans. According to press reports, the FBI reportedly seized the blackboard because, rightfully in their view, it contained classified information. Not all of the incidents were this dramatic, but there were cases where each side had to tread carefully.[34]

Corruption

It might be thought that the topic of corruption is so obvious it does not need comment or, closer to the mark, that it was always assumed to be present but rarely was it so obvious to foreign visitors that it would hit them between the eyeballs. And that was generally true. But there was one instance that was so plain and so glaring that there could be no doubt, and much to the credit of CRDF Global's program on Basic Research and Higher Education (BRHE) in Russia, it led promptly to remedial action.

In its first, main phase, the BRHE program's hallmark was the creation of sixteen Research and Education Centers, or RECs, at Russian universities. These centers were ambitious attempts, through major programmatic financial support, to promote the concept of the modern research university in Russia. The RECs were to do this with generous funding from both the United States and Russia. Almost all the initial BRHE centers competed successfully for continued funding in the second phase subject to the results of in-depth evaluations and site visits by mixed Russian-US expert teams. All but one: Kuban' State University in Krasnodar.

Located in sunny southern Russia, the Krasnodar area has a reputation for a certain amount of independence and "exceptionalism," stemming back in part to its legacy of being the home of the famous (to Jews, infamous) Zaporozh'ye Cossacks. The rector was a very dominant figure and accomplished scientist. We began to suspect that something was wrong, however, when we noticed that some people who appeared to be his relatives showed up in administrative positions at the generously funded REC. Maxim Frank-Kamentskii was a member of the team of visitors. "And his daughter," Frank-Kamenetskii said, "his daughter, his wife, all relatives. . . . No money was given to anybody who didn't belong to his extended family—his in-laws and his direct relatives."

As part of the site visit, there were presentations from the Russian side. Loren Graham and Kamenetskii were the US experts on the visiting committee. At one of the lecture topics, recalled Kamenetskii, "his daughter was [presenting] 'How Fashion Influences Climate.' How fashion influence climate? Not how climate influences fashion, but how fashion influence climate! When she talked, the whole committee was just lying in the floor, rolling with laughter. We could not stop laughing! Laughter. It was something absolutely unbelievable."[35]

On the way back to Moscow from Krasnodar, the visiting committee was undecided about whether that talk alone should disqualify the university from receiving funds for the REC. When they expressed concern about it to the rector, he even denied that the speaker was his daughter. This response, as well as further examination of the financial trail, led to the unanimous decision of the program's binational governing council that the grant should be canceled and the university disqualified from receiving any further support under the program.

Kuban' State University was not the only Russian institution we encountered at which there were administrative or financial problems. Our financial management systems—and those of other US programs—were designed to anticipate and resolve them. This case, however, was truly extraordinary. It was the only one I ever heard of in which, in addition to gross conflicts of financial interest, and largely because of them, the institution's scientific integrity was called into question by its own actions.

Notes

1. See chapter 1.
2. Interview with Karl Western, May 3, 2016.
3. Ibid.
4. Interview with Siegfried Hecker, January 28, 2016.

5. Interview with Eileen Malloy, January 6, 2016.
6. Interview with Rita Colwell, June 9, 2016.
7. Malloy interview.
8. Ibid.
9. My undergraduate professor, Frederick C. Barghoorn of Yale University, often spoke of what he called the "mirror image hypothesis" of Soviet foreign policy: that the Soviets attributed to America the same motives that they themselves held or would be expected to hold under similar circumstances. I have found this concept extremely fruitful not only in the study of Soviet foreign policy, but in life in general.

The ways in which I encountered this lack of understanding were often amusing. I've already described my challenges in the early 1970s in attempting to explain the concept of travelers' checks to Soviet scientists who had just arrived in the United States. Not only did it tax their understanding of my broken Russian, but each time it was an exercise in bridging an enormous cultural gap, as nothing remotely similar existed in the Soviet Union. As far as I know, none of our guests ever went penniless or starved, so these repeated encounters can be counted as brilliant successes in increasing mutual understanding.

10. Interview with Tengiz Tsertsvadze, December 7, 2015.
11. Tsertsvadze interview.
12. Interview with F. Gray Handley, April 20, 2016.
13. In the late 1990s, one seismic detection network in the mountains of Kyrgyzstan detected a Chinese nuclear test in nearly northwestern China.
14. *Bakshish* is a term of apparently Persian origin that Clara Barton once delicately defined as "a gift of money which an Oriental expects and demands for the most trifling service." In the United States, we might call it "bribery," but especially in Eurasia and the Middle East it has deep cultural and historical grounding.
15. Western interview.
16. Ibid.
17. Ibid.
18. Interview with Kip Thorne, October 6, 2015.
19. Interview with Glenn Schweitzer, February 26, 2016.
20. See Shklovskii's discussion of the Lake Sevan conference in chapter 5.
21. Interview with Boris Shklovskii, January 13, 2016.
22. Shklovskii interview.
23. See chapter 5, "Initial Engagement."
24. The RAS Institute of Molecular Genetics had its origins in the late 1950s as a laboratory in the Soviet Ministry of Atomic Energy's Kurchatov Institute of Atomic Physics that was a haven for Soviet geneticists who had managed to survive the decimation of that field by Trofim Lysenko. In 1978, it was transferred to the Soviet Academy of Sciences and established as an institute.
25. Interview with Maxim Frank-Kamenetskii, October 19, 2015.
26. See chapter 3.
27. Thorne interview.
28. Interview with Yaroslav Yatskiv, December 15, 2016.
29. Interview with Moisiei Kaganov, October 16, 2015.
30. But there were often surprises, such as the occasional approval of visits by Soviet laser physicists to Los Alamos National Laboratory. Presumably the folks at Los Alamos wanted to know what the Soviets were up to.

31. This identification came in the form of a coded phrase,, "Donkey Chipmunk," in unclassified State Department cables that we did routinely see.

32. Thorne interview.

33. The reader interested in pursuing this issue further may wish to read Audra Wolfe's fascinating book (Wolfe 2013), which focuses on some of the more bizarre and frightening early intelligence programs of the CIA.

34. Interview with John Zimmerman, September 30, 2015.

35. Frank-Kamenetskii interview.

10

ON THE NATURE OF SCIENCE IN THE FORMER SOVIET UNION

Introduction

A central assumption in the world of international scientific cooperation is "complementarity": that each party brings something unique to the team that the other one lacks, such as experimental versus theoretical expertise, insights and methods from other disciplines, special data banks that are unique to one of the parties, and so forth. In fact, this complementarity is not only a central assumption of international scientific cooperation but of science itself. Increasingly, we understand that progress in the advancement of knowledge is impossible without the cross-fertilization of disciplines, methods, and simply the different ways in which different individuals perceive and think about the world around them.

But what about complementarity in international cooperation? Is it important only because Scientist A is a theorist who happens to live in Country A_1, and Scientist B is an experimentalist who happens to live in Country B_1? That is certainly one way of looking at it, and an extreme advocate of the view that science knows no boundaries, and that there is one and only one commonly used language of science, might conclude that this is indeed the case.

To me, this is an unsatisfying and essentially reductionist argument, yet it is very widespread in the scientific community. It ignores the relationships between the way science is conducted and the way it is structured and funded in different countries, its historical context, and even the relationship between science and language. Do some cultures tend to understand how we come to know things in different ways? Did the empiricism of modern science arise in a vacuum? Could David Hume have formulated the same philosophy of extreme empiricism if he lived in a traditional society rather than in the most industrially and technologically advanced country in the world at the time?

It is certainly true that mathematics, as the lingua franca of science, is precise and unwavering in its characteristics. But is mathematics itself taught and comprehended in the same way in all countries, at all times? Is the way in which scientific researchers use mathematics universally the same? Are scientists from some countries more likely to rely on different areas of mathematics, such as computational mathematics, than others, and why?

And what about the structure of science itself? Have the facts that scientific research in the United States is linked closely with universities, and that industry funds far more of research and development than the government, led to different approaches to scientific research, and perhaps different outcomes, than in the former Soviet Union (FSU), where scientific research was strictly separated from higher education and where research and development was overwhelmingly government funded and linked at the hip with the needs of the military?

Based on my personal exposure to the science systems of the United States and the FSU, my admittedly impressionistic sense has been that there really are differences in the way individuals from these very different contexts do, and think about, science.

To be sure, one glaring difference, historically, has been the ease of access of American and Western scientists in general to sophisticated experimental equipment and computers. As a Soviet physicist told Sig Hecker, "You guys are lazy; you just use the number-crunching computers. . . . We have to think!" I have heard that, too, many times. Does this mean, then, that the American style of science, or at least physics, is more inductive, and the Soviet or Russian style more deductive? That would certainly be a gross oversimplification.

Yet I have continued to wonder whether there are some strands of the interaction between culture and science that might be worth looking into, and accordingly in my interviews for this book, when the opportunity arose, I raised the question to see what would happen. In this chapter, I report the findings. Their nature is that of opinion, not rigorous philosophical reasoning and analysis, but given the extensive experience of those whom I interviewed across scientific communities and cultures, they are hard to ignore.

In addition, the process of scientific reform in the countries of the FSU has been of special interest since the fall of communism in 1991. Even before then, it was well understood by many in that country and foreign observers as well that the old, top-down, academy-dominated, military-oriented scientific research system had to change. The Soviet system's collapse in 1991 offered an opportunity to launch efforts at science reform. Several of my discussants made some interesting observations about this process of institutional change, particularly in Georgia and Ukraine, and I have included them here as well along with some of my personal observations about science reform in Russia, which I have followed closely for many years.

Before delving into the interviews, however, it may help the reader who is unfamiliar with the particular context of Soviet science to take a brief detour in the following section. Those who are familiar with the subject will no doubt find my exposition superficial, and they may wish to skip this section and proceed directly to the next, where I present new, firsthand testimony from my discussants.

A Short Digression on Soviet Science

Throughout Russian history, with very few exceptions, science has served the needs of the state. More specifically, science was first introduced into Russia to strengthen its military defenses: German cannons, under Tsar Aleksey I in the seventeenth century and Dutch shipbuilding under Peter the Great in the late seventeenth and early eighteenth centuries. These technologies helped to catapult Russia to world empire status through its constant battles with the Swedes, Poles, Balts, Turks, and other enemies. Toward the end of his reign, in 1724 Peter established the St. Petersburg Academy of Sciences—now the Russian Academy of Sciences—to cultivate the scientific knowledge and principles on which these military technologies rested. After she assumed the throne in 1762, Catherine the Great, who was fascinated by the European Enlightenment, vigorously supported the academy, which became a magnet for numerous distinguished foreign scientists and the best of Russia's budding scientific community.

Over the next century and a half, Russian science flourished with the support of the state as well as private philanthropists. During the Soviet period, especially after the advent of Stalin to power, the historical pattern of science and technology in the service of military needs was sharply intensified. It is generally conservatively estimated that three-quarters of Soviet science's financial support came from the military.

The Soviet science system, in addition, was characterized by a sharp differentiation between research and education. As noted above, under Stalin, research was concentrated to an extent not seen before into institutes under the USSR Academy of Sciences, as well as some institutes in industrial "branch" ministries, where applied research as well as advanced research on nuclear and other weapons was conducted. Universities largely became pedagogical institutions, with the exception of universities in the two capitals, Moscow and St. Petersburg. Budget priorities for research funding and procurement of scientific equipment—much of which was fabricated in-house by talented technicians—was tilted heavily in the academy's favor. This radical bifurcation of research from education was unusual in scientifically advanced countries. During the Soviet period, however, due to generous funding for military-related research and the huge number of people involved in it, it seemed to function reasonably well, since the institutes themselves essentially took charge of advanced education for promising students graduating from the universities.

The system, however, was enormously inefficient. Graham and Dezhina write that "according to one GOSPLAN [the USSR State Planning Commission] official, the average Soviet scientist was four times less productive than the average U.S. scientist."[1] This inefficiency was a direct consequence of the source and nature of science funding and management in the USSR. Financial support, as already discussed, was distributed noncompetitively, in a top-down manner, to academy and industrial institutes, and from that level by the institute director to research groups and individuals. Merit-based peer review as a method of evaluating and funding scientific research was completely unknown. Instead, the command economy–type system of research management, reflecting the broader, military-oriented command economy of the entire country, bred favoritism and corruption as well as much good science. Lack of access to Western markets for modern scientific equipment meant that even the best Soviet scientists had to improvise with instruments of their own design, which, though ingenious, were usually inferior and less efficient than those used by their counterparts in the West. In applied research in particular, copying and reverse engineering of Western processes and designs was widespread.

This being said, however, there was certainly excellent basic, non-military research carried out in the academy institutes, especially in fields such as physics and mathematics, but also in others. Indeed, the relatively deprived, equipment-poor environment in which they worked yielded insights and approaches that differed markedly from those in other countries, laying the foundation in many fields for complementarity of research approaches that were the key to many successful scientific collaborations.

Counterintuitively, in this top-down system there was also a certain freedom to choose research topics, especially in basic research. Block funding of institutes, laboratories, and groups enabled some top scientists to work on basic research problems unrelated to any particular project. This flexibility in basic research—within certain boundaries, to be sure—was much cherished by Soviet scientists. Indeed, many scientists who remembered and thrived during these times reminisce wistfully about them today. In part, this was also because the system treated scientists very well, especially senior ones and above all members of the Academy of Sciences. Salaries were decent, and often came accompanied by decent apartments at no cost, access to high-quality and imported goods and foods, special rest homes for vacation, and more. To be a Soviet scientist, for many, was one of the greatest distinctions the Soviet Union could confer on a talented individual, and for many it also was a refuge from crass political pressures, where merit really did have some meaning. In the words of a drinking song recalled by Roald Sagdeev, "Only Physics Makes Sense."[2] It is fair to say that while in other advanced countries, the best minds went into careers in a broad range of fields, in the Soviet Union they tended disproportionately to go into science.

I hope that this brief gloss on Soviet science may be helpful to the general reader. The direct testimonies I gleaned from my interviews, however, provide real depth and texture.

In Their Words

On Styles of Science

In physics, at least, it would seem that there is no particular Russian style. There are, however, certainly different styles of physics in Russia, and these are in fact institutionalized into what seem uniquely to call "schools." The "schools" are perhaps the most enduring hallmark of the hard sciences in Russia (and Ukraine), harkening back beyond the Soviet period to the German "schools" of the "grand old men" of physics in the nineteenth century.

The Landau School, named for the great Jewish physicist Lev Davidovich Landau, is the best known of them. It is important to understand that the Landau School is a school without walls; it is not to be confused with the Landau Institute of Theoretical Physics of the Russian Academy of Sciences in Moscow, which was named for Landau after his untimely death in

1968 from injuries sustained in a car accident in 1962. In fact, the Landau School had its origins not in Russia, but in Ukraine, at the Khar'kiv Institute of Physics and Technology—earlier called the Ukrainian Institute of Physics and Technology. Landau later moved to Moscow, where he headed the Theoretical Department of the Soviet Academy of Sciences's Institute of Physical Problems, now the Kapitsa Institute of Physical Problems, in Moscow. He was arrested during Stalin's Great Purge in 1938 for comparing Stalin's regime to Hitler's dictatorship and released, a year later, only after Peter Kapitsa wrote a personal letter to Stalin.

Yet while the Landau School is a school without brick and mortar, it is so institutionalized, in fact, that it even awards certificates of membership. Boris Shklovskii proudly showed me his Landau certificate, which holds a place of honor on his office wall in Minneapolis. As Roald Sagdeev describes in detail in his book, there is (or was) a standard series of exams called the "Landau minimum" that one had to pass in order to become a member in good standing of the Landau School.[3] To be a member of the Landau School was, and still is, a coveted honor. In general, the scientific "schools" in Russia are so revered that in modern-day appeals about the plight of Russian science under the Putin regime, it often seems as if they, and not the deeply institutionalized and invested Russian Academy of Sciences, are the real objects of concern in the Russian scientific community.

Cosmologist Alexander Vilenkin of Tufts University, who studied at the University of Khar'kiv prior to immigrating to the United States, described the Landau School's challenging norms and culture:

> Landau's students kind of adopted his style, which is that when you start giving a seminar, you have a five-minute window to prove that you're not an idiot. And you're making your talk [directly] to Landau. If nobody else understands, it's unimportant. If Landau understands, then it's, "Well, OK, this guy is good," and that is your goal. So it's a very different style from seminars here where it's basically an advertisement of your work, not necessarily like in Russian seminars, when they ask you to give a talk and it's a two-hour talk. Basically in your introduction you tell what it is about, but then you actually do the derivations on the board. Here, nobody wants to see your derivations. They want to see what it is that you did, and if they think it's interesting then they will look at your paper."[4]

Of styles in Russian physics, and the "schools," Kip Thorne said:

> In physics, at least in theoretical physics, I think it probably did not have much to do with the particular culture or philosophical traditions people came from.

It had to do with individuals. In theoretical physics, it's true that really big breakthroughs usually come from individuals, and from individuals who usually stand out above most everyone else in terms of creativity and insight. The whole community is crucial to the regular forward march of science but the flashes, the huge flashes of discovery in theory usually come from individuals.

And in Russia in that era, in the twentieth century, you had Lev Landau who had [a] huge impact both as a mentor and in his research. On the astrophysics side, you had [Yakov] Zel'dovich and you had [Vitaly] Ginzburg. These were people who had great intuition. Zel'dovich functioned almost entirely on intuition plus his amazing ability to take a complicated problem and build a very simple mathematical model that incorporated crucial physics in order to see what was going on. He was unique and used to be great at that. You had other people who made their huge breakthroughs come through calculations rather than intuition.... [There are] different styles.... [Subrahmanyan] Chandrasekhar had little intuition but huge computational abilities—well, enough intuition to know what to calculate. In Russia, Zel'dovich was singularly intuitive, Ginzburg was somewhat the same, Landau was more analytic but had great intuition. But I think it really had to do with these particular individuals and where they got what they had, it's hard to say. But there is what is called the Landau School, and neither Zel'dovich nor Ginzburg came from the Landau School. They in some sense had their own schools.[5]

However, to say that the differences in Russian physics were more the result of individual styles rather than cultural influence is not, Thorne clarified, to say that there are no cultural boundaries in the field:

There are those cultural differences. Certainly, the boundaries, the barriers, that you had to deal with in collaborating [with] the Soviet Union were there. I was able to surmount them because I was dealing with Zel'dovich, and then because later I tried to lubricate things with Braginsky with an interuniversity collaboration, and because my friends kept the CIA and FBI off my back. Those boundaries, barriers, were there; you just had to deal with them. But there were no boundaries or barriers in my field, in the sense that we were all working on more or less the same problems and we had an enormous respect for each other between the Soviet Union and the West. And we were influencing each other....

The communication boundaries, barriers, were big. There was always a language barrier. Most American scientists didn't really know what was going on in Soviet science until the Russian journals were translated. There was roughly a nine-month delay for the translations in the '60s and '70s; for those few of us who could read some Russian, fortunately we would get the information earlier. Zel'dovich, for example, would send a shortened version of articles, sometimes in English but usually in Russian, and then thanks to Zel'dovich I received all the major Russian journals. I arranged for his group to get *The Astrophysical Journal*—the most influential of the Western astrophysics journals.[6]

Thorne's remarks also point to another factor in the development of specific styles of Soviet science, in particular, that cut across all fields, isolation from world science. It was both a disadvantage and, in some senses, an advantage. It was a massive disadvantage in all fields of science because the extreme difficulty in receiving foreign scientific journals set the Soviet scientific establishment at a distance from the rapid development of scientific knowledge in the rest of the world. Counterintuitively, however, this isolation was also seen, perhaps with some black humor, as an advantage in that it facilitated the separate development of unique scientific approaches, much like the evolution of unique species in the Galapagos Islands. Boris Shklovskii explained:

> Probably half of my education came from international and American journals. We would get a journal perhaps nine months to a year later after its publication date. So in this sense, we were at a disadvantage.... But actually, at the end of the day, at least for me personally and some of my colleagues, it became an advantage, because... we were not rushing to keep up with today's scientific fashion. Sometimes we were able to think deeper and slower and invent something original. And so it turned out that what we did was quite original work, and later, of course, acknowledged by the Western community, and this was because of the semi-isolation.[7]

A very different but extremely important perspective on a specific style of science in the Soviet Union involves not the method of doing research, but the organic link in some fields between science and education. One of those fields is mathematics. The tradition of mathematical "Olympiads" is immensely popular and geographically widespread throughout the territory of the FSU. Mathematical prodigies are indeed at the heart of science throughout the region, and it may even be argued that the strong focus on mathematical skills is perhaps *the* defining characteristic of Russian and Soviet science in general. Geometer Marjorie Senechal commented about what she considered unique about Russian mathematics:

> I can only speak about my own impressions, but which is that Russian mathematics, one thing that has made it—at least the mathematics I know about—different from elsewhere, especially French, but also American, is that it's understandable. The reason that it's understandable is that they try to really explain it so that you can understand it, instead of intimidating you by building up these forests of abstractions. Part of the reason for that, though—and this is when the culture comes in—is that there's a long tradition, and this everybody agrees on, of the very, very best mathematicians writing for schoolchildren.
>
> Those mathematics books for children had been properly translated, even before I went there. I knew them; I owned some of them. They were written for elementary school students or junior high, but they were wonderful even

for college students. They explained the main ideas, and how can we use them and how we can represent them . . . they were just magical! And all those books, I didn't realize back then, were written by world-famous mathematicians. This tradition is very, very strong, and this is, I think, what made all the difference.

They weren't just talking to kids one way and to their colleagues another way; this is how they talk. I asked my colleagues there about this. They said that mathematicians feel it is their duty. . . . There wasn't a separation between the universities and the schools the way there is in the US. Even though the universities were not research universities in the way that many are here, nevertheless, there was a continuity through the high schools, the universities, and the research institutes that was very, very strong. . . . Everyone I talked to felt that that was naturally what they do because you want to bring young people into this field; you want them to understand it.[8]

Even in spite of the artificial, damaging and politically motivated wedge driven between research and universities by Stalin in the early 1930s, this linkage between science and primary and secondary education is extraordinarily deep, ingrained in the culture itself, and endured throughout the Soviet period to the present day. Even the very notion of "schools"—of great teachers and brilliant, elite students—bears the telltale markings of this fundamental association. While the institutions of Soviet science may have evolved in such a way as to strain this link in advanced research, in the culture of Soviet science education they were probably stronger than anywhere else in the world.

No discussion of the relationship between institutions and science in the Soviet Union would be complete without mention of the ubiquitous role of secrecy and political control on the conduct of Soviet science. Two of my discussants had specific comments on this topic. They are less revelations of anything new than personal and direct clarifications of practices that are generally well-known but not often documented.

First, Revaz Solomonia on sending scientific articles abroad for publication. Reflecting on recent Russian government policy and practice under Vladimir Putin, Solomonia said:

> Maybe they are moving back to Stalin's time, when it was a closed system. . . . During the Soviet Union, if we were sending a paper abroad, we had to wait for a decision from a special committee to allow you to send the paper, right? It was a really ridiculous situation. You had to write a letter to this committee; I cannot remember what this organization was called, and you had to say, "My paper doesn't contain any new information in it." If your paper does not contain any new information, why are you publishing it? Who would be interested in it? Maybe they are moving back to Stalin's time, I do not know.[9]

I heard this story from many former Soviet scientists. It was also common knowledge, in the Soviet period, that each scientific delegation contained either an intelligence official or an informant. Alexander Ruzmaikin and Kip Thorne discussed this explicitly when I spoke with them:

RUZMAIKIN: But we knew how the system worked. The KGB put people, their officers, directly into the delegations. Because in delegations in which Soviet scientists go to meetings in Warsaw, Paris or London, a so-called scientist would pretend to be a scientist, but he's an officer. That was easy to identify; we knew who worked in science and especially in the field, you go and see a foreign face so you tried to avoid this guy. And the second way: Some scientists worked for the KGB.... I'll give you a name in particular, like a professor of Moscow State University, Dr.Sc. T [name omitted]. You knew his name maybe?

THORNE: I knew T.

RUZMAIKIN: T. He really was a Major in the KGB.

THORNE: Yes, he was also the minder. He was really the weak physicist in the delegation. But he was obvious.... I didn't know that about him, but it was clear that he was the one reporting to the KGB.

RUZMAIKIN: No, no, there is a different level. He actually had a rank like KGB major, which was a very high rank, because it's much higher the rank compared to the army, say, or navy. And the other [way was that] people just reported, voluntarily or under pressure, to the KGB. One of them was known as I. [name omitted].

THORNE: Yes, again I. ...

RUZMAIKIN: Prof. I. was known to report to the KGB, but he did it, I think, from jealousy. If he didn't like somebody, he handed in a report on him, because we all were obliged to write a report about any visits abroad.... Some people like I. and others reported even personal things about foreign scientists and even their own colleagues; scientists like Rashid Sunyayev got in trouble once because one of his colleagues reported that he had a long dinner with two Americans. That got him into big trouble. I don't know who it was. Zel'dovich saved him, but ...

THORNE: So I should remark that I. was one of the people who informed on Braginskiy after the Copenhagen meeting which we discussed earlier.[10] Not the only one, but ...

RUZMAIKIN: I see. So they were driven even by jealousy, because I. was an extremely jealous person. He considered that he's a genius, an unrecognized genius who didn't accomplish anything.[11]

Ruzmaikin pointed out, however, that the practice of filing reports on foreign scientific visits was not only restricted to the Soviet Union. He found the trip reports required of him by NASA, as an employee of the Jet Propulsion Lab in Pasadena, California, to be an equal, and perhaps a greater, challenge:

> It was like a protocol. Just like we do it now with NASA. I'm invited, say, to Europe next spring. So I file a travel request with NASA. It goes through JPL, JPL goes through NASA, and NASA after three months will give me approval. But if I already bought a ticket and paid my registration fee, then I'm in trouble. So the same was in Russia. You have a travel request; for that, you need an invitation from the organizing committee to prove that you are going to present your talk and so you have to [submit a request]. And then you're approved. But the initial stage of approval is similar to what NASA has now, even worse than the KGB; I will tell you, now if you talk about NASA, it's worse. I knew how to get around the KGB rules, but I don't know how to get around the NASA rules. They're so strict, so hard. So when you return, you have to write what they call a "report," your account of your visit—in a positive sense. Usually you say I attended this meeting, I presented my talk, and I met say Prof. Thorne, we discussed that and that subject. Or some people mention we went out for dinner and so on. . . . In the Soviet Union, I personally tried to avoid personal things; I knew they catch the personal things when they read your report; so you report only the science.[12]

On Science and Language

Another cultural factor mentioned by Thorne was the language barrier. Alexander Ruzmaikin, who, valued his collaboration with British scientists for its value in helping learn how to write scientific papers in English,[13] sees the particular kind of language used to this day in Russian physics papers as a discrete form of communication: "Because the Russian language is implicative, if you understand what I mean. They write something assuming that you understand what they mean. . . . Sometimes my Russian friends send me Russian scientific articles to read. . . . Now I can hardly read them, because you have to guess what they actually mean."[14]

When the opportunity arose, I pursued this line of inquiry about the relationship between science and language in subsequent interviews. When I spoke with botanist Peter Raven, it suddenly exploded. Not only, Raven pointed out, do linguistic differences influence science in his field; they enrich it immeasurably. I take the liberty of quoting his comments at length:

> Biological classification is a large and difficult subject largely because there are so many kinds of organisms, by no means always sharply distinct from one

another. We know very little about many of them, and our decisions about how to classify them must necessarily consist of a series of hypotheses of varying strength, based on the information on which the decisions were made.

In the early 1960s, I worked with what is called folk taxonomy—the kind of system that is used to classify plants and other kinds of organisms by people who do not have the ability to record their decisions and observations in writing. All of their names for organisms exist only as parts of their spoken language. Actually, such groups of people treat the names they use very differently from the rather formal way that we do: if the plants, for example, are useful, they may have many names for different races of what we'd consider a single species, and if they're not, the opposite may well be true.

Language is the way we communicate: understanding its principles can be complex. Languages are not codes but different ways of expressing ideas and names; they reflect deeper principles of thought.... Computers can help us to understand the differences and similarities, and in a field like mine, allow us to deal with many more parts and the relationships between them than would be possible in any other way. When it comes to giving names to different kinds of organisms, we can do that but still retain in our databases the information that informed our conclusions—computers allow us to do that.

One of the major points of decision in biological taxonomy is how many species are represented in a given group of specimens, or individuals. And Russian and Chinese and Americans might consider that there were different numbers of species, given a set of information, because to some extent they look at the nature of species differently. Typological classifications express the degree of difference; evolutionary classifications express relationship by evolutionary descent. Breaks between species, discontinuities in the patterns of variation can be very sharp or gradual, and the differences between the entities themselves can either be obvious or rather small and trivial.... The observations that a scientist has actually made in reaching such decisions are of enduring importance regardless of what the scientist decided to do about them in a particular classification system. Lord knows I'm no philosopher, but . . . you never can look at the outside world and assume you know or understand it completely....

People like Goethe or Alexander von Humboldt led us to think about the world of nature holistically, but we always have a tendency to break it up so we can communicate about its particular parts. If our only classification systems are verbal, we can "carry" only a limited number of names, because we have to store them in our brains or lose them. When we write about them in books, we can have as many names as we want. When we get to databases stored in computers, we can not only do that but [also] record as many features as we like about the individual kinds of plants and even the exact measurements that went into the conclusions about the nature of particular features. Philosophically, we have to deal with both the Cosmos, the unified, integrated natural world that von Humboldt was one of the first to understand and explain, and with the individual elements in that Cosmos. In our time, the English philosopher James Lovelock has used the concept of Gaia, Mother Earth, to make some of the same points about unity.[15]

While I am not personally familiar with the literature on language and science, as a Russian speaker, I have often wondered about whether the language itself may have had some impact on how Russian (and other Slavophone) scientists may see the world and do their science. For the time being, I leave these thoughts for others to ponder. They are but one strand of the rich insights I have gained from my interviews, but I feel that they must be important beyond the purposes of this book.

On Science Reform

Each country that was part of the FSU has sought to deal in one way or another with the Stalinist model's legacy of top-down science funding, extremely vertical organization of research institutes, and sharp bifurcation of research from education. Each country has sought to introduce one degree of reform or another, with varying success. The old system may have worked reasonably well to support the needs of the military imperial autocracies and dictatorships of the nineteenth and twentieth centuries based on the Russian Empire, but it was poorly suited to the task of adapting to market-based economies in which technological innovation runs ahead of and responds to civilian and military needs alike.

An early internal indictment of the Soviet science system was made in 1988 by Roald Sagdeev, then director of the Soviet Academy's Space Research Institute and a Soviet scientist in transition to becoming an American scientist. In his article, "Science and Perestroika: A Long Way to Go,"[16] he argued that Soviet science was beset by "bureaucratic dinosaurs," the erosion of scientific standards, poor scientific and technological planning, lack of access to powerful computers, and international isolation. "Reform" was just in its early stages, but, he said, was incomplete.[17] In a way, Sagdeev's 1988 vision of reform itself was a prisoner of the times and did not even come close to addressing what might be changed in the transition to a market economy, no doubt because that eventuality was not on anyone's mind at the time. But it was soon to happen.

In the following discussion, we begin by looking at the way science reform has played out since 1991 in Russia. Since I was unable to discuss these issues with Russians with current information on the topic in interviews for this book, and since this is a very long and involved story that has yet to be thoroughly documented, I will summarize it only briefly in this space.

The two key post-1991 science reform issues in Russia have been, first, the role of the Russian Academy of Sciences (RAS), the principal heir of the massive Academy of Sciences of the USSR (ASUSSR). The RAS inherited some six hundred institutes from the ASUSSR and extensive associated property holdings, making it one of the largest property owners in Russia (by area). When domestic research funding virtually disappeared in the early 1990s, many institutes started renting out their unused working space to commercial enterprises, primarily small businesses. In the atmosphere of wild, intensive property acquisition during the early Boris Yel'tsin years, the RAS's property holdings became an extremely tempting target for both private investors and the state. In addition, and more importantly, it became clear to many that the academy's top-down structure and methods, suitable to a command economy, were incompatible with an emerging market environment where innovation and competition were key to economic growth and viability.

A closely related problem was the gap between research and education. Advanced training in science, in the Soviet system, was under the purview of academy institutes, not universities, with one or two exceptions. Western models of educational excellence at research universities became objects of keen interest, and projects appeared that aspired to boost Russian universities into the Top 100 universities in the world. In addition, struggling Russian universities, as well as some academy institutes, heard the spectacular success stories of some US universities with "technology transfer" and imagined that technological innovation could be the magic solution to their deep financial problems. It was not.[18] Nevertheless, the view that research and education needed to be reunited in an invigorated university system became fairly widespread, particularly in government (but certainly not in the academy).

One of the earliest voices in the early 1990s to articulate the need for change was that of Boris G. Saltykov, an economist who became the Russian Federation's first minister of science and technology policy under President Boris Yel'tsin from 1991 to 1996. Saltykov advocated a radical restructuring of Russian science, with a significant reduction in the role of the greatest bureaucratic dinosaur of them all, the massive Russian Academy of Sciences. In a 1997 *Nature* article, Saltykov explained his vision to Western readers. He called for deep reforms to break up the country's centralized research network, both in the academy sector and that of the industrial branch ministries, and to strengthen the integration of research and education in the universities.[19]

In this crusade, Saltykov, a soft-spoken man of refined—even noble, some might say—manners and speech, made few friends and many enemies. In 1997, Saltykov's ministry was demoted to the status of a state committee (which had been the customary level of representation in the Soviet period) and further reform efforts were effectively put on hold for the time being.

Another important but half-hearted step toward structural reform was taken when, in late 1992, the government established the Russian Foundation for Basic Research (RFBR), which launched a program of competitive grant funding.[20] The RFBR, however, was chronically and severely underfunded and remained a minor source of support for Russian scientists.

Saltykov paid the political price for his advocacy of serious structural reform, but subsequent ministers of education and science (the functions became combined), especially under the leadership of Andrey A. Fursenko,[21] went even further. They called for new charters for the academy, stripping the academy of its hundreds of research institutes, increasing researchers' salaries and correspondingly cutting the number of researchers, and, most importantly, refocusing the country's effort by strengthening the role of universities in science—integrating research and education and effectively introducing the concept of the modern research university into Russia. All of this eventually came to pass, but not always in a way that reflected well on the Russian government. There were protests by scientists and studied ignorance by the authorities. And in the end, many questions still remain.[22]

In 2017, a dramatic development in this story in Russia was the forced resignation of RAS president Vladimir Fortov, who was elected to that position shortly before the 2013 events in which the academy's institutes were transferred to a new government organization, the Federal Agency for Scientific Organizations (FASO). This was evidently a very ugly affair. In early 2017, the RAS was scheduled to hold its next presidential election. Even in Soviet times, one of the academy's most important prerogatives was to have its elections held independently, secretly, and with no government interference. This time, however, was different. Fortov campaigned for reelection on a platform[23] of reasserting the academy's role in the formulation of basic research policy and setting priorities for its former institutes.

Fortov's demonstration of independence, relatively meek as it was, proved intolerable to the government. According to an account published in the independent occasional publication read widely in the Russian scientific community, *Trotskiy variant*, Fortov was summoned peremptorily to the Kremlin, threatened there with a criminal investigation into financial irregularities in his institute (the Joint Institute for High Temperatures), returned to the academy, and on March 22, 2017, resigned as president and designated one of its vice presidents, Valery Kozlov, as acting president.[24]

The final nail in the coffin of the RAS institutes was put in place in May 2018, when the Russian government abolished the Ministry of Education and Science and replaced it with two ministries, a Ministry of Education and a Ministry of Science and Higher Education. FASO, with the former academy institutes under its aegis, was subsumed in the new Ministry of Science and Higher Education. FASO's head, Mikhail Kotyukov,[25] was named the new minister of science and higher education.

The humiliation of the once august and powerful Russian Academy of Sciences, established by Peter the Great, was complete. This, for the time being, is the main result of science reform in Russia. While many keen observers of Russian science agree that reform of the academy and the integration of research and education at Russian universities were important goals in the post-Soviet period, the sordid manner in which it was done will leave a black mark on Russian science that will be hard to erase.

The result, ironically, is probably closer to what Stalin would have preferred ninety years ago when he tore research out of the unreliable universities and relocated it into the Soviet Academy of Sciences. From the standpoint of a strictly authoritarian government, the situation is now arguably better. Instead of merely sequestering advanced research in a semi-independent and bothersome academy of sciences, it is now directly subjugated to direct government control through the new Ministry of Science and Higher Education and the traditional industrial branch ministries. This is hardly an arrangement conducive to growing a knowledge economy, in which government funding of research typically takes second place to that by private companies. It is, however, an excellent arrangement for maximizing control and minimizing creativity and innovation.

Moving on from Russia, the two countries that I did visit for this book in December 2015, Georgia and Ukraine, are literally at the opposite ends of the spectrum when it comes to reform of the science system.

The first former Soviet countries to implement far-reaching reform were actually Lithuania, Latvia, and Estonia, which had done so much earlier. The pattern in those countries was to strip the academies of sciences of their research institutes and to distribute them to universities, as was also being done at the time in some neighboring countries formerly in the

Soviet sphere of influence, notably Poland. In 2005–06, Georgia joined their ranks, dispersing the Georgian Academy of Sciences's research institutes to universities and other institutions, and converting the academy itself into an advisory and honorific body. The Georgian academy's budget was slashed and remains a mere shadow of its former self. Today, if you ask scientists in Georgia how that all came out, the first thing you will hear is that the country's science budget was eviscerated and that scientific research in Georgia is on the ropes. To be sure, this crisis is due in part to the fact that Georgia has itself been beset by war and hostile annexation, not to mention economic turmoil.

But when in December 2015 I met with scientists from Georgian universities, in particular, I heard a very different story, and one that I found rather surprising. Revaz Solomonia, director of the Institute of Chemical Biology at the Ilia State University (formerly a technical university) in Tbilisi, told me that being associated with the university gives him "more opportunity to have better science." He explained:

> It creates a difference with the research system because . . . research and education were separated completely. The better science was in the research institutes under the Academy of Sciences, but the researchers at the academy were not teaching at all, and they were the best. And the universities were mainly teaching universities and the scientists were not as good as at the research institutes. And now we have the opposite situation, and now the science is better. . . . Because . . . after the reforms of 2005, the research institutes cannot educate students. How can a research institute educate students? . . .
>
> First of all, we should take one idea, the indisputable idea that research and teaching are inseparable from each other. I think this is a basic idea. I cannot say anything about Russia because I have not been to Russia for maybe twenty-five years or so, and I have no idea what is going on there, but what was the situation in Tbilisi? It was the old Soviet system. At the time, for example, at our Institute of Physiology, we had our council that was allowed to award PhD degrees. For this reason, we had a lot of PhD students. . . .
>
> In 2004–05, there was a decision [in Georgia] to take all the institutions out of the Academy of Science. The budget was extremely low; I cannot explain how low the budget was. The institutes were nearly empty, and only the ones who had some international funding were surviving. It was not completely like this here because still I was saying that the government should give two more years to PhD students who had already started their PhDs. Practically they had put all these people on the street, and the privilege of granting PhDs was given only to the universities. And so now all the universities have a chance to have PhD students, and automatically the better researchers went to the universities. Most research institutes were merged with universities; universities have more money, larger budgets, more young people, and the research is going better.[26]

I pursued this question with another scientist I met at Ilia State University, Zurab Javakishvili, director of the university's Institute of Earth Sciences. When I asked him if his experience reflected Solomonia's, and if in particular the budget was even growing, he confirmed that it was so. Why? "Because it's easy for them to cut funds to science, but very hard to cut funds for education." But then he mentioned something else in passing that I found fascinating: that while at the academy there were a half dozen institutes in various areas of the earth sciences, at the university there is only one single department that covers all of them in a holistic, interdisciplinary manner.[27] This is not only a matter of convenience to accommodate a smaller staff but also a strategy common to universities in general. Thus, the Department of Earth Sciences at the university includes mineralogy, geochemistry, geophysics, magnetism, and other related fields, and students get exposure to many of them and work with faculty on interdisciplinary projects.

From both of these discussions in Tbilisi, it struck me that in this sense, the dismemberment of the Georgian academy resulted not only the transfer of many scientists to universities and to teaching, but also the replacement of a highly vertical, compartmentalized organization of scientific subdisciplines by a more horizontal, integrated approach to the conduct of science itself. In this sense, too, Georgia's science reform appeared to result in an improved atmosphere for science and advanced education in science, as Solomonia also mentioned, not only because of the benefits of associating research with education but also because in the process the very organization of scientific disciplines changed as well. This is a point that is not usually made with regard to science reform in the FSU, and it would be interesting to see to what extent it has been or might in the future be validated in Russia or other former Soviet countries where reform is much further along.

At the other end of the reform spectrum has been Ukraine. Until December 2015, the Ukrainian government resisted any and all attempts to introduce any meaningful modifications to the traditional Soviet-style system. While some Ukrainian universities did take steps to improve the quality of their research, such as the Kyiv Polytechnic Institute and the independent Kyiv-Mohyla Academy, steps to dismantle or restructure the Academy of Sciences system itself were stymied for years. At least two laws on science reform since 1991 had been vetoed by Ukrainian presidents. The main obstacle was the venerable National Academy of Sciences of Ukraine (NASU), headed by its equally venerable president, Borys Yevhenovych Paton. Indeed, according to legend, the two are exactly the same age, as Paton was born on the same day as NASU was founded in 1918. Konstantyn Yushchenko, a senior scientist at NASU's E. O. Paton Institute of Electric Welding, told me that at the age of ninety-seven, Paton was still vigorous, swimming daily and personally involved in academy decisions.[28] The deep respect in which he is held—he is said to be the only scientist in the Soviet Union to whom a statue was erected during his lifetime—and his iron hold over the academy's institutes, have made him a formidable force to be reckoned with in post-Soviet Ukraine. I also heard it said in December 2015 that meaningful science reform in Ukraine would not happen during his lifetime—which as of this writing spans a century.

While the National Academy of Sciences of Ukraine was holding on to its coveted privileges, Ukrainian science was suffering serious losses, and not only because of the ravages of war, economic turmoil, and corruption. The fundamental problem, said Yaroslav Yatskiv, director of the Main Astronomical Observatory in Kyiv and a NASU officer, is that there "is some lack of understanding at the highest levels of government that countries like Ukraine cannot exist without the development of science and technology. We are not Russia, we are not Kazakhstan, we are not Turkmenistan and so on, and we are not even Romania. And only through high technology can we find our place in the world."[29]

But whether because of poverty, or misplaced priorities, or even a sense that Ukrainian science in its present state was not significantly contributing to national economic development, the government did not pursue the "knowledge economy" path after the Soviet Union's demise.

The result, Yatskiv said, was a massive brain drain from Ukraine. While I do not know of any studies comparing the exodus of scientists from Ukraine and Russia, Yatskiv described to me a dire situation:

> All my young colleagues . . . are already overseas. Take for example Sergei Bolotin at the Goddard Space Flight Center—a great scientist who left and now works in the United States of America; take M. Mishchenko, who used to work at the Main Astronomical Observatory, now with the Goddard Institute for Space Studies in New York, a famous specialist there in atmospheric science, and so on. In Germany, some of my former employees are working on laser ranging, and even in Australia. From my small department of fifteen scientists, I have been left with only three to five people, and they all have special reasons to stay in Ukraine. They have been invited to work overseas, but do not want to leave their family, or do not want to go for other reasons.

> This is the greatest problem that our government does not understand, that we have had until now so many talented young people, we are still a talented nation, but their talent cannot be applied here. They are forced to leave to find self-realization and to feed their families.

Of those who have emigrated, Yatskiv said, "half of them have come back, the other half have not; they come back to visit. If there was a divorce, they come back."[30]

From Yatskiv's account, it was not clear to me that the scientific emigration, both "external" to other countries, and "internal" to fields other than science, was that much different in scope or in kind to what was going on in Russia, except for one thing: the Ukrainian government did not appear to be doing anything about it, whereas the Russian government had at least made some motions in the direction of reform and even launched special programs, including one funded privately by Dmitriy Zimin,[31] to attract émigré Russian scientists back from abroad. It was all of a piece with the complete lack of any sign that the government was ready to contemplate science reform in Ukraine, much less allocate more resources to it.

While I was in Ukraine in December 2015, the Ukrainian Rada (parliament) introduced a new bill with a comprehensive law on science reform. It involved a major shift in funding strategy, with 80 percent of government research funds to be allocated competitively through a new peer-review agency. It was widely expected that it would suffer the same fate as its predecessors and be vetoed by the Ukrainian president, Petro Poroshenko. I spoke about this with Boris Movchan, founder of the International Center of Electron-Beam Technologies and reportedly a close friend of Paton, about his assessment of the need for reform and the prospects of the new law. I found him refreshingly open to the idea, noting that the French had managed to alter their government's structures for support of science and allowing that this might also be possible in Ukraine.

At the same time, however, Movchan pointed to another aspect of Western support for science that is often overlooked in discussions of science reform in postcommunist countries, which focus on issues like the presence or absence of powerful academies of sciences and the role of universities, but not on another vital point—the role of the private sector in supporting scientific research:

> What I know from the literature is that as a rule some kinds of new beginnings occur in the private sector. Small companies. Then, they go out on to the market. With half the money from the government and half the money private.

And I consider that for our technology, that would be ideal. . . . What you're talking about I completely agree should be, because now, through our system of supporting sciences, nothing remains. But we don't have private capital. . . . Competition is needed. Independence is needed but also private capital. It is necessary that our oligarchs, unpleasant people, [get involved]; the money that they earn is going out of Ukraine. This is nonsense that there is no such [private] money [for scientific research] . . . But we live now in a complicated time for Ukraine. Look what goes on every day. This is nonsense. . . . This is my opinion; I don't know, maybe I hold too sharp opinions, but in general this is impermissible, how we live now, and for that matter in science. Therefore, the reform that you speak about is necessary and will begin.[32]

From everything else I had heard during those two weeks, that prediction seemed uncharacteristically optimistic, but then again, perhaps Movchan, due to his friendship with Paton, knew something that everyone else didn't know. That discussion took place on December 17. And indeed, one week later, President Poroshenko signed the science reform law, a move that caught many by surprise.[33]

Implementation of the Ukrainian science reform law, as of 2017-18, was very slow but proceeding apace. Through an abundance of caution and concern about transparency, a number of oversight and evaluation committees were formed, with each step along the way subject to approval by the prime minister. An international Expert Committee met several times and was in the process of selecting competitively a Science Committee to oversee proposal evaluation. Chairing the Expert Committee was Dr. Sergiy Ryabchenko, a distinguished physicist, former chief government science official in Ukraine, and member of the National Academy of Sciences of Ukraine.[34] His role in the reform implementation was a hopeful sign. There were two very important unknowns: how much money would actually be devoted to the competitive review process, and especially what would be the role and position of the National Academy of Sciences of Ukraine. To my knowledge, as of this writing, these are still open questions.

There are deeper reasons than bureaucratic complexity, however, to the slow pace of progress. First of all, Ukraine is still a country at war with the Russian-supported insurgents in the East. Secondly, according to Ryabchenko,[35] "arranging the furniture" of science institutions and committees is secondary to answering the fundamental question, "Why should taxpayers' money be spent on science?" On this, he says, "the new Law says not a single word." What is needed, he adds, is a thorough discussion of the role of science and innovation in the Ukrainian economy, which has yet to be undertaken. A 2016 *Peer Review of the Ukrainian Research and Innovation System* carried out under the auspices of the European Union's Directorate for Research and Innovation concluded that "the country needs to 'innovate its path to growth' with a cross-governmental STI Strategy that is backed by adequate tools. This will require a cross-government effort that involves the intellectual, material and financial assets of the country. Ukraine must place research and innovation high on its political and policy agenda. The Strategy should be developed and implemented to exploit the potential of STI for growth and societal wellbeing."[36] It is a long path, but it appears that the right questions are being asked.

Thus, while the Georgian model of science reform was to relocate the former academy institutes under university auspices, and the Ukrainian model has so far left the academy institutes intact, in Russia the former vast empire of academy institutes is, for the first time ever, now under direct ministerial control. It will be interesting to see how each of these variants plays out in terms of scientific productivity, innovation, and education in the years ahead.

Notes

1. Graham and Dezhina 2008, 13.
2. Sagdeev 1994, 32.
3. Ibid., 39–47.
4. Interview with Alexander Vilenkin, October 15, 2015.
5. Interview with Kip Thorne, October 6, 2015.
6. Kip Thorne, commenting during interview with Alexander Ruzmaikin and Joan Feynman, October 6, 2015.
7. Interview with Boris Shklovskii, January 13, 2016.
8. Interview with Marjorie Senechal, May 10, 2016.
9. Interview with Revaz Solomonia, December 8, 2015.
10. See chapter 6, "Kip Thorne: Gravitational Wave Detection."
11. Discussion between Thorne and Ruzmaikin, in Ruzmaikin-Feynman interview.
12. Ruzmaikin, in Ruzmaikin-Feynman interview.
13. See chapter 6.
14. Ruzmaikin, in Ruzmaikin-Feynman interview.
15. Interview with Peter Raven, February 23, 2016.
16. Sagdeev 1988.
17. Ibid.
18. In this, there were many things, in their enthusiasm, that they did not or chose not to understand: that most US university technology transfer offices fail to generate surplus revenue and only a few score big successes; that university-based innovation requires a strong basic research foundation, which was lacking at nearly all Russian universities; that it takes a long time, as well as creative and enterprising work to form partnerships with private industry; that

it requires a sound legal basis not only in intellectual property legislation but also in rule of law in general and an independent judiciary; and not least of all, that it requires a thriving private industrial sector and a domestic market for high-technology products.

19. Saltykov 1997.
20. On the RFBR and the introduction of investigator-initiated, competitive research grants, see also chapter 4, "Major New Programs Come – and Go," and chapter 7, "Impact on Scientific Infrastructure."
21. In 2014, Fursenko, as well as fifteen other Russian government officials with close ties to Putin, was named on the US government sanctions list in response to the annexation of Crimea. See Stone, "Embattled President Seeks New Path for Russian Academy," 2014.
22. This story is too long to reproduce here in any detail, but it could be part of a good dissertation entitled, for example, "How science institutions fail." There was much good reporting on these developments, of which the following are only a sample: Balzer 2015, "New Ministry for Russian Science" (May 25, 2018), "Russian Roulette" (July 4, 2013), Schiermeier 2013, and Stone 2013 and 2018.
23. Fortov 2017. This was a glossy, multipage campaign document, probably the first of its kind in the history of the Russian Academy of Sciences.
24. Demina and Mesyats 2017. The article's authenticity is strengthened by its coauthorship by Gennadiy Mesyats, a senior vice president of the Russian Academy of Sciences.
25. "Siberian Financial Whizz-Kid Appointed to Lead Russian Science" 2013.
26. Solomonia interview.
27. Interview with Zurab Javakishvili, December 8, 2015.
28. Interview with Konstantyn Yushchenko, December 15, 2015. E. O. Paton, the institute's founder, was Borys Paton's father. As of March 2019, Paton had passed the 100-year mark.
29. Interview with Yaroslav Yatskiv, December 15, 2015.
30. Ibid.
31. Zimin's project was through his private foundation, Dynasty, which among other things provided generous fellowships for outstanding young physicists as well as prizes for émigré scientists who decided to return to Russia. He was forced to shut down Dynasty and leave the country in a bizarre episode of the Putin government's crusade against foreign-funded nongovernmental organizations. It was alleged that because he funded Dynasty from personal offshore accounts, it was subject to the foreign NGO restrictions and since it had not registered as such, the claim was that its operations were illegal. See "Russian Science Foundation Shuts Down after Being Branded 'Foreign Agent.'" I had heard that the real reason for this action was that Dynasty had sponsored a symposium at which some speakers spoke less than adoringly of the Putin government's policies.
32. Interview with Boris Movchan, December 17, 2015.
33. "President: Ukrainian Science Should Become Modern, Efficient and Innovative." The text of the law can be found at "Pro naukovu i naukovo-tekhnichnu diyal'nist'" ["On Scientific and Scientific-Technical Activity"].
34. Ryabchenko was the first chairman of the post-Soviet State Committee of Science and Technology of Ukraine in the early 1990s.
35. Personal email messages from Sergiy Ryabchenko dated April 23 and 24, 2017.
36. *Peer Review of the Ukrainian Research and Innovation System* 2016, 8.

11

VIGNETTES

THE BEST AND MOST FUN PART OF RESEARCHING and writing this book was hearing the stories from the folks I interviewed. Most of their stories appeared in previous chapters as they related to one specific theme or another. Others, however, defy categorization. I offer them without comment. I have thrown in a couple of my own favorite tales as well.

Stephen Hawking Goes to Moscow (Kip Thorne)

"So I took [Stephen] Hawking with me [to Moscow] in '72. The Russians had invited him, but they didn't have any idea how to deal with his particular problems. He and I were very close friends, so they asked me to come with him, with him and his wife. . . We spent a lot of time talking with Zel'dovich and Zel'dovich's young collaborator, [Alexei] Starobinsky. Zel'dovich had just realized that spinning black holes should radiate. This was a huge breakthrough and very surprising. Stephen had intense discussions with Starobinsky and Zel'dovich about this, and he went home. Triggered by those discussions, he realized soon thereafter that all black holes would radiate with what is now called Hawking Radiation. The idea—the seed—for Hawking Radiation was his discussions with Zel'dovich and Starobinsky in Moscow on that visit. . . .

Zel'dovich had a tremendously deep physical intuition. He did not understand the mathematics of black holes, he didn't understand relativity very well, but he did understand intuitively that with any rotating object where there's energy available . . . if you were to throw something in, such as particles—this had already been shown by [Roger] Penrose—if you throw a particle in, and it splits apart, one piece goes in the black hole and the other one comes out. The one that comes out can come out with more energy than the original one. There could be amplification, and you're getting the energy from the spin of the black hole. Zel'dovich realized—nobody else had realized this; this was discovered simultaneously by Charlie Misner in the United States—that if you sent waves in, the waves could be amplified, and a wave would go in and basically it would get energy from the spin of the black hole, and it would come out with more energy than it went in with....

Zel'dovich discovered that intuitively. But he realized that this is a general property of rotating bodies. He first thought about it for a rotating electrically conducting sphere, a sphere of steel or iron or copper—that if you sent in electromagnetic waves, they could be amplified. The energy would come from the spin of the sphere, and he could do simple calculations showing that this was so. But he was so deep—he understood the connections between waves and particles so deeply—that he realized then that if what you sent in was a vacuum, which means nothing, the vacuum in the real world has virtual particles, particles that are flashing in and out of existence all the time; you can't get rid of them but they can't become real either. So that if you send in a vacuum [to] a spinning body, including a spinning black hole, the energy of the spin could be fed into those virtual particles and they could become real. And therefore you're just sending in the vacuum—you can't avoid sending in the vacuum—and when the spinning body spins it will turn some of those virtual particles into real particles, and the spinning body will radiate particles.

So Zel'dovich saw this all largely intuitively. He could do simple calculations for spinning electrical metallic bodies; he couldn't do them for black holes. But his young colleague Starobinsky could. Starobinsky was a graduate student at the time, so Zel'dovich told Starobinsky that this was going to happen and Starobinsky did the calculations and showed that it would happen. Then they discussed this with Hawking and me. Zel'dovich had already told me this in private before, based on his intuition; he had not yet had Starobinsky do the calculations. So I knew this from a prior discussion with Zel'dovich.... Zel'dovich hadn't published anything, and I didn't believe it. I made a bet on it. This was my first scientific bet, which I lost to him ... whether spinning black holes will radiate.

So I was there with Hawking when Zel'dovich gave the intuitive argument, which Hawking didn't buy. Then Starobinsky sketched the calculation, and Hawking said, "I will go home and think about it." He went home and thought about it and discovered Hawking Radiation, which is a much deeper phenomenon and a much more surprising phenomenon, that all black holes will radiate. You don't have to have the energy of spin. Hawking was much deeper than any of the rest of us, but the stimulus came from Zel'dovich. It was those kind of interactions, and there were many of them—this was the most spectacular of them. These interactions are crucial for the progress of science, and I had the pleasure of playing a significant role in facilitating and catalyzing them."[1]

The Expedition from Hell; or, Science as a Contact Sport (Julie Brigham-Grette)

"In 2000, our group flew Anchorage to Magadan, and then my Russian colleagues had organized a cargo plane to take us up to Pevek. . . . I had to take all of my money for the project with me [in cash]. . . . I actually flew over to Russia with $70,000 wrapped to my waist. And then I had to take part of that money and put it into rubles to pay for the charter plane to Pevek, about $23,000. We were going from bank to bank for a week because the banks would only let you change US dollars into rubles so much a day. . . . I've got pictures of me with the money and I felt like bank a robber. And actually I had my Russian colleagues to protect me. . . .

But as we waited for the charter to Pevek from Magadan, I knew something was up. The guy who owned the plane ran a firefighting company . . . and he used this plane for that. He had our gear and he kept putting off our trip. I'd say, 'OK, we're going to leave,' and he'd say, 'No, no, no.' Well, he was cooking up a deal, as it turned out. He knew how much gear we had, and he got us out to the Magadan airport which is fifty kilometers from town, and he had it choreographed.

We're all ready to go, and he says, 'Oh, we can't get all this stuff on the plane,' and we said, 'Yeah, right.' I mean he had our stuff for a week and a half, and he then decides he can't get all the stuff in the plane?

He said, 'I'm going to have to make two trips.' That was OK, because we had paid him to take us up, drop us off, and then come back a month later and bring us back, right? That was the two trips. . .

Anyway . . . he flew all of us, all the people and half the gear to Pevek, and he says, 'OK, I'll be back tomorrow with the rest of the stuff.' Well, then it took another five days. Then he shows up with . . . the rest of our stuff, and he sits on the tarmac, and in those days, this was 2000, there was not a lot of fuel around. And of course he's running this firefighting company.

So he says, 'You can't have your gear out of my plane until you give me your jet fuel.' He wanted to basically take the jet fuel that we had purchased for our helicopter flights from Pevek out to the lake and back, and he was going to use it for his own use. . . .

So we sat and swore, we were swearing and yelling at each other in four different languages, and I was sitting in the manager of the airport's office and I kept trying to say, 'If we were in America you'd be in jail, man. You can't do this!' And we went way out on the point near Pevek and I got on my SAT [satellite] phone and I called Simon Stevenson [at the National Science Foundation (NSF)] and next to me was my German partner, Martin Melles, and he's on the phone calling the German people. And Simon, I remember, said, 'Look, this will be a good experiment. We'll get you out of there, we'll figure out how to do this.' That's when he connected with CRDF, and they connected me with Shawn Wheeler, who then negotiated a new contract with a different Russian aviation company while we were in the field.

I remember being on the satellite phone, and Shawn finally said, 'We've got it all figured out, don't worry.' He said, 'When you're ready to come home from Pevek to Magadan, just give this company a call. They'll come and get you.' And I remember saying to him, 'Well, wait a minute, is there a project number or invoice number I need to give him so he knows who we are?' And he goes, 'Don't worry, they know who you are.'

At the end of the expedition, we were coming back from the lake and, remember, I had this thing going on with the helicopter . . . at the end of the expedition. Well, the first helicopter came in, got half of us, brought us back to Pevek, and the helicopter then was delayed because a couple of guys got on that helicopter and drank vodka all the way back to the lake. They got into a fight at the lake over a woman, and we got on the helicopter, got away from these guys, and flew away. Well then, one of those guys went and shot our meteorological station with a gun and shot right through the data logger, and that shot stopped all the measurements. We had $30,000 or $40,000 of equipment out there to measure the meteorology. . . . That expedition was the one from hell, it was amazing.

So, [when] I went to Wrangel Island [later], I also went out to the lake for a day with a box of equipment to fix the met station. So I get out to the met station—I hadn't seen it since I'd left—and I open up the box and sure enough. . . . There was this one place where there's the on-and-off switch, and he had shot right through there, unbeknownst, right through the on/off switch so I had something like forty or fifty wires I had to redo with a brand-new data logger. And the miracle was, I'm on the phone with the guy that built the met station; he's in Fairbanks, Alaska, and I'm on satellite phone with him, I told him what's going on, and he goes, 'OK, get busy, get going, get it going,' and he talked me through how to rewire this thing. I had never done this before in my life. This is complete divine intervention, I'm sure. Because I got it all ready, and he said, OK, now do this, do this, and then listen for the clicks.' We all sat there listening. Click. Click. . . . That system operated continuously until I took it apart in 2009. . . .

That was 2000 when the met station was shot. In 2001, when my Russian colleagues from St. Petersburg went to get my permits to allow me to go to Wrangel Island, there were some problems. The KGB, whatever they were at that time, they had it in their notes somewhere that I had an illegal meteorological station at the lake because of the shooting incident. . . . This was crazy of course, and even the local authorities asked me about this in 2003 when we went back out to do more fieldwork. So I was on their little hit list as being some trespasser, even though we had permits for that met station to be there, we had all the paperwork. The local people don't care what Moscow says, they just, you know, decided that I was a troublemaker. . . .

But, it's worth it. . . . During the 2000 trip, another thing happened. . . . When I gave this charter plane owner the money to pay for our plane ahead of time, I went in, I remember, it was so funny, we went into the storefront in Magadan and I had told my Russian colleague Pavel, I said, 'Pavel, you know, in Alaska, you pay half of it now, and half of it later. That's usually the way we do it.' He said, 'Oh, no, I know, I know this guy, it's no problem.' I remember taking all this money into the storefront that looked abandoned. We walked into the store, nothing on the shelves, there was a room in the back, just like out of a detective movie or something. And there in the back, three guys in suits and tie[s], and they were the owner[s] of the company, and then there was a woman in the corner—I distinctly remember this, I'm not making it up. She had a miniskirt on, big tall boots, long hair, and she was sitting there playing with her hair. . . . and all decked out. It was like a mafia thing. That's what I felt. In the middle of all of this, I'm feeling like this Girl Scout, right?

. . . So we gave them the money and he's now making me a receipt, and he said, 'Is it OK, we'll make you out this receipt in Russian rubles?' I said, 'Yes, fine, I'd expect that. I mean I'm paying you in rubles. It's OK.' And he goes, 'So do you want us to make it for the full amount or should we add something else?' Right? So they wanted to know if I wanted a personal kickback from my receipt, and I said no, just make it for the exact amount of the dollars. . . .

There were also times when they wouldn't let us go fly to the lake, between Pevek and the lake, with the helicopters because the weather was bad at the lake. This was the old manager of the airport. And he said, 'Oh, I'm sorry, the weather is bad there, so you can't go.' So day after day after day we go to the airport. 'Come on,' you know. So, by that time, we pretty much had befriended everybody in the airport and the guy that actually ran the met station through the airport he started talking to my friend and he goes, 'It's not the weather, it's totally fine.'

And I said, 'OK, what do we do? Look, I'll go offer the guy money.'

And sure enough, I offered the guy money. I said, 'One thousand dollars, can you make two flights happen today?'

'No problem!'

I'm in a staircase, I hand the guy the cash . . . he puts the thousand dollars in his coat. Suddenly the weather was perfect, and we flew out there."[2]

[After hearing Brigham-Grette tell this story, I commented facetiously, "I'm thinking of a title for this section, 'Science as a Contact Sport.'" "That's for sure," she said.]

The Reception (Moisiei Kaganov)

"One day before opening of the conference,[3] Galya and I were finalizing some last-minute issues. Suddenly, the room filled with strange people—'strange' for an academic department. Some of them wore military uniforms. Others were in civilian clothes, but we had no doubt that they were military as well. We immediately understood that we were not the main hosts: they were asking questions and we had to answer. Mainly, they wanted to meet with the chief, i.e., with Ilya Mikhailovich Lifshitz. Galya explained that I was in charge of administrative questions. They were only interested in knowing who from the West would be attending the conference. I explained that famous theoretical physicists from different countries agreed to participate in the conference and a large, obviously more representative delegation from the United States. They did not ask directly, but, I think, I guessed correctly: they wanted to know whether any physicists involved in the development of nuclear weapons would attend the conference. I am not sure what they concluded from my answers. They disappeared as unexpectedly as they appeared.

One of them stayed with us. Unlike his colleagues, he introduced himself and said that he was an authorized KGB agent at the Presidium of the Academy of Sciences. He gave us his phone number and suggested to call him if necessary. He also asked whether there would be a reception for the delegates. After learning that it would take place at the Hotel National's restaurant, he asked for an invitation for himself and asked or demanded (probably 'advised' would be a more correct description) to send a few invitations to the employees at the international department of the Academy of Sciences. I understood that, just like him, they were KGB employees. He openly said: 'This is not your last conference. You will probably need to travel abroad. A lot depends on them. . . . '

The fact that *gebeshniks* [slang for KGB agents] were given a few tickets bothered me. I knew that many of our and foreign participants knew each other well. We were supposed to serve alcohol at the reception. It makes people talk. Who knew what people would discuss? It would not be a good idea for KGB ears to hear them. I warned some participants who were my close friends. At the reception, I was keeping an eye on the intruders. Soon I stopped worrying. It was obvious: they were only interested in delicious food and good drinks. By the time participants continued their conversations holding cups of coffee, all of those people who were of concern to me had left the reception. They were not interested in anything else: the food and drinks were gone."[4]

She Wanted to Touch Me (Moisiei Kaganov)

"Foreign physicists seldom visited Kharkov. Therefore, I remember two visits, that I will be describing, very well.

The first foreign scientists who wanted to meet theoretical physicists from Kharkov, I. M. [Ilya Mikhailovich] Lifshitz and his students, Mark Ya. Azbel and myself, were American physicists, Richard Bozorth and [a man named] Goldman. At that time, our research in the field of electron theory of metals and magnetism were widely quoted. Therefore, we were not surprised that they wanted to meet us.

We knew who Richard Bozorth was very well. He was a recognized authority in the field of magnetism physics. We were not familiar with Goldman, I did not research his work (tried now, but unsuccessfully). During our conversation, we learned that Goldman was a chief scientist at the Ford Company Research Labs. Mr. Goldman came with his wife, a typical American woman, as I now understand.

They stayed at the Intourist Hotel. They were not allowed to visit UPTI. We did not get permission to invite them to our homes either. We met and spoke at the hotel (I do not remember where, specifically; in the room or in the lobby), at the hotel restaurant, and while taking walks in Kharkov. There were three of us from the Soviet side: those whom they wanted to meet. Neither I nor Ilya Mikhailovich spoke English at that time. We spoke through Mark. He was a participant in our discussions and our interpreter. Our leadership, most likely because they were sure that wiretapping worked reliably at the hotel, did not assign anyone to monitor us. We were without a monitor at all times.

We spoke about physics, about our results, and plans. I don't remember well, most likely. I also remember attempts, especially by Mr. Goldman and his wife, to switch the subject of our conversation and discuss the condition of Jews in the Soviet Union. They probably did not quite understand what could be said, and where.

Mr. Bozorth was more careful. While taking a walk together, he started speaking about Jewish scientists not being able to travel abroad. We, of course, confirmed it, but said that sometimes there may be other reasons, such as participation in classified work. He nodded with understanding.

I must say that I did not like what the Goldmans were saying in the beginning, especially him, but her as well. For example, Goldman was trying to convince us that we saw Americans as fat rich people with a big cigar. I asked him why he came to this conclusion. He replied:

'That is how we are portrayed in *Krokodil*.'⁵

'But I don't read *Krokodil*!' I objected.

Goldman's wife was saying that, in her opinion, the worst part of life of the Soviet Jews was that Jewish children could not attend Jewish religious schools—*heders*. I replied that there were other, much worse things, such as when someone is unable to obtain a graduate degree, find work as a professional, and so on. Our conversation continued and I said, 'If the absence of *heders* seriously concerned Soviet Jews, I would have been the first one to be concerned.'

'Why?'

'Because I am the great-great-grandson of Rabbi Itzhak Elchanan Spector!' I said the full name of my great-great-grandfather.

My guest's reaction astounded me: she asked my permission to touch me and later told me how much Rabbi Spector is respected, in particular among Jewish people in New York. She spoke about some schools named in his honor. I have to admit that in twenty years living in the United States, I have not researched this subject. Although I did learn something over the years. Information comes to you even if you are not looking for it."⁶

International Women's Day (Marjorie Senechal)

"'March 8 is Women's Day, a legal holiday,' I wrote to my mother from Moscow. 'This is one of the many cute cards that is on sale now, all with flowers somewhere on them. We hope March 8 finds you well and happy, and enjoying an early spring! Alas, here it is −30° C again.'

I spent the 1978–79 academic year working in Moscow in the Soviet Academy of Science's Institute of Crystallography. I'd been corresponding with a scientist there for several years, and when I heard about the exchange program between our nations' respective academies, I applied for it. Friends were horrified. The Cold War was raging, and Afghanistan rumbled in the background. But scientists understand each other, just like generals do. I flew to Moscow, family in tow, early in October. The first snow had fallen the night before; women in wool headscarves were sweeping the airport runways with birch brooms.

None of us spoke Russian well when we arrived; this was immersion. We lived on the fourteenth floor of an academy-owned apartment building with no laundry facilities and an unreliable elevator. It was a cold winter even by Russian standards, plunging to −40° on the C and F scales (they cross there). On weekdays, my daughters and I trudged through the snow to the broad Leninskiy Prospekt. The five-story brick institute sat on the near side, and the girls went to Soviet public schools on the far side, behind a large department store. The underpass was a thriving illegal free market where pensioners sold hard-to-find items like phonebooks, mushrooms, and used toys. Nearing the schools, we ran the ever-watchful Grandmother Gauntlet. In this country of working mothers, bundled bescarved grandmothers shopped, cooked, herded their charges, and bossed everyone in sight: Put on your hat! Button up your children!

At the institute, I was supposed to be escorted to my office every day, but after a few months, the guards waved me on. I couldn't stray in any case: the doors along the corridors were always closed. Was I politically untouchable?

But the office was a friendly place. I shared it with three crystallographers: Valentina, Marina, and the professor I'd come to work with. We exchanged language lessons and took tea breaks together. Colleagues stopped by, some to talk shop, some for a haircut (Marina ran a business on the side). Scientists understand each other. My work took new directions.

I also tried to work with a professor from Moscow State University. He was admired in the West and I had used him as a contact on my application. But this was one scientist I never understood. He arrived late for our appointments at the institute without excuses or apologies. I was, I soon surmised, to write papers for him, not with him. I held my tongue, as I thought befit a guest, until the February afternoon he showed up two weeks late. Suddenly the spirit of the grandmothers possessed me. 'How dare you!' I yelled in Russian. 'Get out of here and don't come back!' 'Take some Valium' Valentina whispered; wherever had she found it? But she was as proud as she was worried. The next morning, I was untouchable no more: doors opened wide and people greeted me cheerily, 'Hi! How's it going?'

International Women's Day, with roots in suffrage, labor, and the Russian Revolution, became a national holiday in Russia in 1918, and is still one today. . . . On 7 March I was fêted, along with the institute's female scientists, lab technicians, librarians, office staff, and custodians. I still have the large copper medal, unprofessionally engraved in the institute lab. '8 марта'—8 March—it says on one side, the lab initials and the year on the other. The once-pink ribbon loops through a hole at the top. Maybe they gave medals to all of us, or maybe I earned it for throwing the professor out of the institute."[7]

AIDS! AIDS! AIDS! (Karl Western)

"In the early HIV days, the East Germans and others were saying that the HIV virus had been manufactured in the US and that it was a biological weapon that had been implanted on, you know, the underprivileged . . . here, you know—the gays, the minority populations—and that we planted it in sub-Saharan Africa and Haiti. There was a US-Soviet health meeting that was scheduled, and there was a health agenda and it was going to be held in Washington—the CDC had the lead—because the State Department was in Washington, not Atlanta. The head of the Soviet delegation was Viktor Zhdanov. He was one of their leading virologists, and he was one of the early people that got into HIV. So, Zhdanov and his delegation were coming and both sides were instructed not to discuss HIV by the State Department, because the Soviets were throwing out this information and . . . anyway, it was taken off the table by the diplomats. You can talk about influenza, you can talk about hepatitis, you can talk about polio, but you can't talk about HIV.

Zhdanov was a very outspoken person. . . . His limousine pulled up in front of Building 31 [the location of NIH's Fogarty International Center]. He got out, and he told his NIH counterpart host, who was Dr. John LaMontagne, who became our deputy director, as soon as he got out, he said, 'I'm an old man, I have to go to the bathroom.'

So John showed him where the bathroom was, and I trailed along, and Zhdanov was up at the urinal, and he was shouting, 'AIDS! AIDS! AIDS! There, I've said it. Now we can talk about it.' We did talk about it. And he got into lots of trouble back in the Soviet Union; I think he was fired as laboratory head or forced into premature retirement."[8]

How I Missed Dinner with Jonas Salk (Eileen Malloy)

"One of my jobs, when I was serving as a consular officer at the US embassy in Moscow, was to be the liaison with all the spouses of American citizens who'd been denied Soviet exit visas to join their spouse in the United States. One of them was a young lady and the supposed reason for the denial of her exit visa was that her father was a physicist. In those days, that was enough. She spoke English very well and we got along very well, so we would see each other from time to time. She called me one day and said, 'I need your help. I have a family friend coming to visit and I need you to drive him from the train station to my parent's house.' Her parents lived out in the outskirts of town in one of the high-rise apartment buildings. She didn't tell me who it was. . . . So, I agreed to do that, and I met her at the train station and then this couple got off the train. It was Jonas Salk and his wife, Françoise Gilot, who used to be Picasso's mistress. . . .

It was fall, and you know in the fall in Moscow it's rainy, snowy, slushy horrible weather. We get in my car—a brand-new Volvo. We're about halfway out of town, and my car dies. And here I'm going to have dinner with Jonas Salk. And it doesn't get any better than this. And I am working at the embassy, you know, it was just really cool—and my car dies. So I'm outside and it's raining and slushy . . . you're up to twelve inches of slush and standing out there, and I've got the hood up. I don't know anything about cars. I'm trying to figure out what is wrong.

And all the sudden there's this man at my shoulder saying, 'Well, let me see if I can help you.' I look over and it's Dr. Salk, and what flashed through my head is he's going to die of pneumonia and I'm going to be the cause.

So I said, 'No, no, you have to say in the car. Stay in the car.'

And so he gets in the car and this Russian girl gets out, and she says, 'Well, maybe we should flag down a cab.' So she flags a cab.

They all get in the cab, she gives me the address, they go. Eventually, I get the car going, but I cannot find the apartment building. You know those forests of unmarked buildings? Never found the building. Never made it to dinner and I missed it. But what so impressed me is he could come and go—he had that relationship. He could call on people that I couldn't see, but they could see him. . . .

Well, he finally got her exit permission to depart the USSR. He was the one that broke the logjam and she went to work for him in La Jolla [at] his institute."[9]

The Phantom Award (Eileen Malloy)

"Mike Joyce was . . . DCM [deputy chief of mission] in Embassy Moscow when [Ambassador] Jim Collins was there, but in 1982, he was science counselor and I worked for him in the Science Section. . . .

He was always looking for ways that we could interact with the scientific community, because, of course, this was very much hands-off in those days, and one day he read a Soviet newspaper—I forget which one—and there was this little article, way in the back pages, about a group of Soviet scientists who had received a prestigious American scientific award. It listed their names.

He came to me and said, 'We must hold a reception for these scientists.'

I asked, 'How do we do that?' because there were no addresses or even what cities these people lived in listed in the newspaper article.

'We'll talk to MinScience,' he said.[10]

It was my job to go to MinScience and explain we wanted their help in contacting these people, and for months they put me off and put me off. They wouldn't give us ways of contacting the scientists. They wouldn't take the invitations to forward to the scientists. Mike was frustrated with me. I was frustrated. So finally I went in person to the Ministry—I actually think it was the Soviet Academy of Sciences. . . .

This would have been in 1982. And I showed up in the midst of a drenching downpour in a cab, and for some reason I had my then three-year-old child with me. I don't know why she ended up at the embassy that day. I left her in the cab, which was good, because otherwise the cabby would've abandoned me; he's left with this American child in the back of his car.

I ran inside the Academy of Science building and said, 'Look, I need to understand what the issue is here.'

The Soviet Academy staffer looked at me and said, 'You put us in a very difficult position.'

It turned out that the reported American award didn't exist and that the Soviet scientists didn't exist. The Academy of Science staff thought we were trying to embarrass them. It was a made-up story. There was no award—I mean, we never thought to check to see if there was any such American award. It didn't exist. It was a piece of propaganda [disseminated] by the Soviets. And here, we were trying to force them. . . . It was in the Soviet newspaper. . . . They had just made this thing up. . . .

And these people didn't exist. They had never gotten an award and the award didn't exist. And here, for months, I'd been hammering them. We were going to host a reception for these people. They were so angry with us, because they thought we were trying to expose them. . . . I remember the look on the Soviet Academy staffer's face, and it was just pure gold, when you could get a Soviet to admit that."[11]

How I Almost Got Thrown Out of Russia (Gerson Sher)

When I became CRDF's founding president in 1995 at the request of NSF director Neal Lane, I was confronted with a new problem. I had devoted my entire career to promoting civilian basic research cooperation between the United States and the former Soviet Union (and eastern Europe). I had never dealt with, and never imagined I would have anything to do with, defense research or researchers. But CRDF's authorizing legislation gave it a mission that spanned the two areas, in the sense that it was to promote US-FSU science cooperation in general, it was also to address the issue of redirecting FSU scientists who had worked on weapons of mass destruction to peaceful civilian research through international cooperation.

One of my favorite horror stories about my adventures in Russia comes from CRDF's diligent efforts to ensure that it did indeed meet the second objective of redirecting former weapons scientists. For my first visit in Russia as CRDF's president in the fall of 1995, my main task was to secure the cooperation of the Russian government with CRDF's core activity, the Cooperative Grant Program, or CGP. We modeled CGP after the NSF's cooperative research grant programs, in which there would be an American principal investigator (PI), a Russian (or other former Soviet) PI, a joint statement of work, and a cost estimate outlining the costs in each country. In addition, to help us fulfill our nonproliferation mission, we wanted to be sure that we attracted proposals involving, on the Russian side, former weapons of mass destruction scientists. Finally, it was important to us that the proposals be truly "grass-roots" and that they did not have to pass through the FSU scientists' institutions.

We had a fine idea on paper, but it was extremely naive: To identify these proposals, to which we would give some form of priority in the grant selection process, the Russian scientists would indicate on the cover page of the proposal if they were former weapons of mass destruction scientists. This would be a convenient and transparent way for us to identify such proposals and to ensure that they received some kind of priority attention in accordance with our dual mission. There was a problem with this approach, however: in terms of Russian legislation, it was highly illegal.

So I arrived in Moscow in November 1995 armed with a fine portfolio of briefing materials from the CRDF staff and sat down with Zurab Yakobashvili, the head of the Russian Ministry of Science and Technology Policy's International Department. We had already met and we liked each other. From my previous experience in Russia I well understood that Yakobashvili either worked for or reported to the Federal Security Service (FSB)—the successor to the KGB; this was very familiar territory to me, and it did not bother me at all. As a manager, I even welcomed the presence of Soviet Academy or Ministry staffers who (may or may not have) held parallel positions with the KGB or FSB for one simple reason: they got things done. Things worked. If problems came up, as they inevitably did, they fixed them. It came with the territory.

But never in my career had I worked on this particular type of program with Russia, with investigator-initiated, competitively reviewed cooperative research proposals, because this was the very first time it was being done.

Zurab listened politely to my opening remarks and vision of the program, and then he quietly took me into an adjoining private room. There, he politely but firmly informed me, and not at all hostilely, that if I persisted in advocating this particular program concept he would have to have me declared a persona non grata in Russia and expelled immediately from the country. We looked at each other for a few seconds, and then it occurred to me why he had presented me with this ultimatum. How stupid! We were asking Russian scientists to willfully violate state secrecy laws by disclosing that they worked on weapons of mass destruction. This was so obvious that I did not even have to say anything else to Zurab except to propose that I get together with his staff and work out an appropriate and mutually acceptable way for the Russian participants to identify themselves without getting sent to Siberia.

The solution was not difficult to find, but it required that we at CRDF relent on one of our preconceptions, that the Russian institutes would not be involved in any way with proposal submission. It quickly became obvious that if we invited scientists who had worked in the weapons sector to self-identify in the body of the proposal *if they wished*, and if we agreed that all proposals receive the imprimatur of their institute, then the Russian secrecy laws would not be compromised. Since the institutes were responsible for ensuring that all employees comply with secrecy requirements, their imprimatur on the proposals would be a satisfactory way of achieving both sides' objectives.

When I came back to the United States and explained the problem and the solution I had agreed to in Moscow, I was actually glad that Yakobashvili had read me the riot act so that our staff and our State Department funders would understand the gravity of the situation and why we had to relax one of what we had considered our cardinal principles. The program was a smashing success and remains the core instrument of CRDF Global to support cooperative research projects, its primary (at least in an ordinal sense) mission.

The Ruckus at the Holiday Inn (Gerson Sher)

One day in October 1993, my dear friend Marianna Voevodskaya, who was part of a group of some eighteen Russian PhD scientists who were in Washington, DC, for in-depth training on processing research proposals by peer review in preparation for the International Science Foundation's first Long-Term Research Grants Program competition, came into my office in Georgetown with a very serious look on her face. She sat down opposite my desk and told me, "Gerson, I have something to tell." This was the first of many times over the next several years that Marianna, who became something of a personal confidante, presented me with unpleasant information.

From Marianna's grave manner I sensed that something was terribly wrong. "What's happened?" I asked.

Marianna then proceeded to tell me the following story. The previous night, many of the men in the delegation (sixteen out of the eighteen guests) got together in the room of Sergey Mashko in the Holiday Inn on Wisconsin Avenue in Georgetown. Mashko had been designated by our Moscow Office, which was our main operational unit, to supervise the "scientific secretaries," as they were called. He came from the infamous Biopreparat Institute, which had long been the Soviet Union's main offensive biological warfare institute. How that happened was never clear to me as the Washington, DC–based ISF chief operating officer, but Sergey certainly seemed, and was, competent and congenial enough.

So all these male PhD scientists from Moscow crammed themselves into Sergey's hotel room in Georgetown and proceeded to do what Russian males always do in hotel rooms—smoke and drink. And they smoked and drank a lot. Finally, their cigarette smoke had set off the room's smoke alarm and the sprinkler on the ceiling started to spew out water on the bed and all around the room. Now, it being well known that Russian scientists, especially male Russian scientists, can figure out how to deal with any situation, Sergey bravely climbed up on the bed and attempted to disarm the sprinkler. But American technology defeated him. The sprinkler assembly fell apart, and water began to gush from the ceiling pipe with renewed vigor. It flooded the room, it flooded the hall, and as Marianna told the story, there was about a quarter inch of water on the entire floor. The women—Marianna and our mutual friend Naya Smorodinskaya (there may have been someone else, perhaps a spouse or two)—armed themselves with towels and, kneeling on the floor, attempted to mop up the flood.

By this time in Marianna's story, which she related with great gravity and earnestness, I was rolling on the floor. I couldn't stop laughing. Of course it was a very serious matter and I was worried that the hotel was going to dun us with a huge bill for damages, and possibly worse. But as it turned out, I never heard from the hotel management, which may have been just as amused as I was. In any case, I was much relieved, since I did not want to explain to George Soros why he had to pay thousands of dollars in damages to the Holiday Inn due to the carousing of "my" employees for whose visit in the United States I was responsible.

Elsewhere in this book, I have touched on the subject of science and culture. If by 1993, I had not made that association, this incident at the Holiday Inn in Georgetown brought it all home to me in living Technicolor. It was a key learning moment in my ongoing education about US-Russia science cooperation.

Notes

1. Interview with Kip Thorne, October 6, 2015.
2. Interview with Julie Brigham-Grette, December 2, 2015.
3. This was the International Conference on Solid State Theory, also discussed by Kaganov in the section of chapter 9 titled, "Intelligence Gathering and Secrecy."
4. Interview with Moisei Kaganov, October 16, 2015.

5. *Krokodil* was a Soviet humor magazine.
6. Kaganov interview.
7. Marjorie Senechal, "8 marta 1979: Women's Day in the Soviet Union," OUPblog: Oxford University Press's Academic Insights for the Thinking World, accessed June 16, 2016, http://blog.oup.com/2014/03/international-womens-day-soviet-union/.
8. Interview with Karl Western, May 3, 2016.
9. Interview with Eileen Malloy, January 6, 2016.
10. In Soviet times, there was actually no Ministry of Science. This was US embassy shorthand. The reference could have been either to the Academy of Sciences of the USSR, technically a nongovernmental body, or to the USSR State Committee on Science and Technology, a body attached to the USSR Council of Ministers. The more likely reference in this case would have been to the Soviet Academy of Sciences, which was the more prestigious and far larger organization.
11. Malloy interview.

PART III

CONCLUSION: SO WHAT?

12

WHAT TO MAKE OF IT ALL?

Was It All Worth It?

How do you answer such a question? In this book, I looked at the experiences of sixty-two people—scientists, diplomats, government officials, program managers, entrepreneurs—and asked them to evaluate their initial expectations in light of their actual experience. Not everyone answered that question directly, but what struck me was that nobody answered it with an outright negative—that no, it wasn't worth the effort, the money, or the time. Of all the individuals I interviewed and who answered that question directly, the only somewhat less than positive answer I received, unexpectedly, was from Nobel Laureate Roald Hoffmann. He told me that in spite of all visits to the Soviet Union, Russia, and Ukraine; in spite of all his many friends and followers there; in spite of the fact that his involvement with that part of the world began as a graduate student and continued for the next some sixty years almost without interruption, that it had no real impact on his scientific work. What interested him, he said, was "the people. I found them interesting." Is that a positive evaluation or a negative one? Surely, Hoffmann's repeated visits and contacts were a conduit for the informal flow of information, which was important for the scientists he visited; he also got some talented postdoctoral researchers out of the deal. And presumably the fact that a Nobel Laureate would regularly shuttle back and forth between the two countries is itself a mark in the plus column.

What about the collective impact, the global impact in terms of President Dwight D. Eisenhower's goal of mutual understanding, Secretary of State Henry Kissinger's goal of restraining Soviet behavior, George Soros's goal of saving the best of the Soviet scientific community for science after 1991, or the broadly shared goals of preventing nuclear war and hostile states and groups from obtaining means of mass destruction, stimulating institutional change, and the obvious goal of the advancement of knowledge?

Loren Graham, who has devoted more time pondering these questions than most, had this to say to me:

> Of course, I'm prejudiced because I participated in so many of these cooperative projects, but I would maintain it was worth it. . . . [If] you've got a hard-nosed person looking at this history, some of them are going to say to a person like me, or you, "You know, you were pretty naive, weren't you?" I mean, wasn't it naive?
>
> Former Civilian Research and Development Foundation (CRDF) board chair Gloria Duffy said in 2001 that "the largest scientific aid program in history is going on between the United States and Russia."[1] So some senator could say, "So, the largest scientific program in history went on, and what came out of it? Wasn't that just a lot of wasted money?" I would disagree with that position, but I think that in the court of public opinion, the skeptical senator would probably win the argument. It wouldn't win it with me, but it's a powerful argument. . . .
>
> But there was a time when you could have a powerful counterargument, and it went like this: "With the collapse of the Soviet Union, Russia is gradually, perhaps a little painfully but nonetheless gradually, joining the Western world, and it's just going to be a part of a bigger Europe. It will never be like the United States, just like it will never be like France or Germany or Great Britain. It has its own culture and traditions. But nonetheless it will exist in the greater Europe and the principles of democracy and the rule of law and so forth will work. And we, who are scientists or are in the scientific community, ought to play our role in helping that to happen." Some people will laugh at that now and say, "How naive you were." I don't think it was so naive, there was a time when even Putin said that Russian might join NATO. It sounds incredible now, but he did say it.[2]

I will leave it for the readers to come to their own conclusions about whether the sentiment was naïve. But whether or not it was an accurate reflection of reality, it did represent a powerful intention and aspiration on the part of the US scientific community, and indeed the world scientific community. Though it may not all have worked out the way it was intended, even if the hope was naive, does that mean it was not worthwhile? And what does this mean for the larger enterprise of international scientific cooperation? Must we always be cold, hard-nosed realists, or may we also be driven by our vision of a better world?

In the following sections, I will look at some different ways of summing up the outcomes of the sixty years of scientific cooperation spanned in this book. None is self-sufficient, but together they offer at least a basis for arriving at lessons learned that may benefit other programs of international scientific cooperation, current and future.

I begin with two key issues that were constant objects of contention: the first—who won?—was intensely debated by Western policymakers and analysts in the Soviet period, and the second—cooperation versus assistance—was more pertinent to the period after 1991.

Who Won?

Equality, Reciprocity, and Mutuality of Benefit

The key guiding principles of the original umbrella exchanges agreement of 1958, which remained in effect until 1991, were equality, reciprocity, and mutual benefit. Equality was pretty straightforward, signifying both that the two countries were equal partners to the agreement and entitled to equal treatment, and also that the participants in the exchange programs were peers and to be afforded equal treatment and equal respect in their activities.

In practice, the principle of reciprocity was probably the key operative principle that governed the actual implementation of the programs. Issuance of special visas for scientists and scholars was one of the most important innovations. With the 1958 Lacy-Zaroubin agreement, exchange visitors were issued special visas that identified a formal host organization, the location or locations where the visitor was authorized to reside and travel, and what financial responsibility, if any, the visitor bore. This system, while highly formal and bureaucratic, gave exchange visitors a measure of security and predictability that did not exist before. Three years later, the United States extended this system to other countries with the introduction of the "J-1" exchange visitor visa, which the host organization submitted to the Bureau of Cultural Affairs at the State Department for approval and which then served as the basis for visa issuance by the corresponding US consulate. While I am not certain if this special visa category originated with the US-Soviet exchange programs, it seems likely, since prior to that time, as Richmond attests, such overarching exchange agreements were unknown.

Another innovation was symmetry of a sort with financial support arrangements. Expenses were in principle shared, usually with each side covering expenses within their countries and transportation to the other ("receiving-side pays"), or in some cases with each side responsible for all expenses of their own participants ("sending-side pays"). Of course, since the two currencies were not exchangeable or comparable, it was impossible to determine whether there were equal levels of financial contribution, but it was in fact irrelevant. Each side covered its own expenses, such as they were. In this case the symmetry of financial arrangements—either sending-side or receiving-side pays—generally held during the détente period, since the unconvertible ruble remained a stumbling block to true cost-sharing.

It was the issue of mutuality of benefit, however, that proved by far the most difficult and contentious of all.

The differing priorities and political climates of the two countries made it inevitable that ensuring mutuality of benefit would be a challenge. Indeed, one of the primary reasons for the formality and central focus on diplomatic agreements was so that each side could demonstrate it was getting commensurate rewards from the relationship. In the United States, in particular, the issue of mutuality of benefit tended to become a political football in times of discord, which was often, as there was a common and—in my view and that of others closer to the ground—simplistic perception that the Soviet Union was gaining more from the relationship than the United States. In truth, however, the balance of benefits was more complex. To a great degree, one's assessment of the balance of benefits depended on the yardstick one chose to apply.

To digress briefly, Joan Feynman,[3] while she did not have significant collaborations with Soviet counterparts, spoke to me about the prevailing culture of condescension toward Russian science, especially space science, as late as the early 1990s. Turning to her husband, Alexander Ruzmaikin, during an interview held at Kip Thorne's home in Pasadena, California, she said: "I would say that at the time you came over [1991 or 1992] and before that time, the prejudice against Russian scientist was enormous. The prejudice against Russian space scientists was particularly enormous. It was so enormous that when some of my colleagues saw that I was with him . . . they told me not to marry a Russian, to get rid of him, because Russians are all cheats. This was somebody they didn't know; they didn't know him, and they didn't know me, and that [perception] was sort of normal: "Russians are all cheats!"[4]

For those actors and observers sitting in the national security agencies or in the Congress, mutuality of benefit, as seen from a thirty-thousand-foot perspective, often seemed pretty clear: in science and technology, the argument went, they were the winners because they were "behind" us, while in scholarly exchanges in the humanities, we were the main beneficiaries because their historical and political records were generally closed, while ours were open. What was worse, on the negative side of the balance sheet, was the understanding (which was correct) that the source and beneficiary of most advanced Soviet research was the military. In the 1980s, the US Department of Defense published a more or less annual series of glossy public reports, *Soviet Military Power*, demonstrating how advanced US military technologies were being copied or stolen, in part through programs of scientific cooperation.[5]

Undoubtedly, there were such cases. There were probably many others that resulted from careful Soviet reading of the open US scientific literature, particularly of university basic research that may or may not have been funded by the Defense Department but in the general domain, from informal conversations at scientific conferences, or espionage. Undoubtedly, too, it went in the opposite direction—otherwise, indeed, how would we have known about such cases in the first place? It was also generally known that the US intelligence agencies sought to debrief returning American exchange scientists after their visits in the USSR. While there were probably many Americans like Kip Thorne who refused to comply for reasons of principle,[6] this was an important source of information that others felt that the government had a legitimate reason to collect.

If the question about mutuality of benefit was which side gained more scientifically, in terms of the advancement of knowledge, then the answer was more nuanced and might have hinged on the specific field of science in question. Thus, the National Academy of Sciences' 1977 Kaysen Panel concluded that exchanges in theoretical physics and mathematics,[7] where it judged Soviet work to be at least of equal if not superior quality to that in the United States, were more to the benefit of the United States, while in other areas that depended more on advanced experimental instrumentation were probably to Soviet advantage in light of the latter's lack of access to such equipment. To be sure, there was a strong imbalance in genetics and related fields because of the long-term damage done to Soviet science by Lysenkoism. On the other hand, in the "field" sciences, such as earth and atmospheric science, botany, and zoology, the balance was probably at least equal because it gave both countries' researchers access to unique geographic locales, and in fact the preponderance of these collaborations involved field research in the Soviet Union.

In addition, the nature of scientific research in the Soviet Union and the United States—and indeed in advanced industrial countries in general— was fundamentally different, even despite the common observation that the "language of science" and its methods are universal. Yet on a meta-level, as Loren Graham has suggested, philosophical world-outlooks have also had an important impact. In his *Science and Philosophy in the Soviet Union*,[8] Graham makes the intriguing argument that in certain cases, especially in early Soviet science, fundamental new ideas about the nature of reality, leading to radically new and fruitful scientific paradigms hitherto unrecognized by reigning scientific approaches, may have sprung even from the traditions of Marxist thought itself, which emphasized disruptive change in place of evolutionary change. So, for example, the early Soviet biochemist Alexander Oparin theorized, contrary to the gradualist evolutionary theory of that time, that the origin of life on earth was instead the result of violent, disruptive events such as the sudden injection of energy through lightning or some other sources into the primordial "soup" of chemicals in the world oceans. Graham suggests that this insight—now one of the major scientific paradigms about life's origin—came at least in part from Oparin's understanding of the Hegelian-Marxist dialectic, which emphasized qualitative change rather than (or following on to) quantitative change, occasional revolutionary change disrupting plodding evolutionary change.[9]

There were yet other, more practical reasons why Soviet science often offered unique insights that were unlike those of their Western counterparts, as discussed in chapter 10. These included the strong Russian tradition of schools of science based on the German model; a distinctly Russian tradition of science education, focusing on mathematics and theoretical physics from an early age; and finally, an emphasis on theory and computational mathematics necessitated by the absence of sophisticated, modern scientific instrumentation and computation. Many a knowledgeable Western scientist visiting a reasonably good Soviet scientific laboratory would remark on the Soviets' ingenuity in using what they had, with devices of their own making, to conduct sophisticated experiments.

In areas other than mathematics and theoretical physics, the evidence was far less clear that Soviet science was on par with standards in the West, including in the United States. Thus in gross terms, one may say there was an imbalance. But such a sweeping statement would ignore unique approaches, achievements, institutes, and individuals from the Soviet Union that were genuine attractors for American scientists, especially in the field sciences—earth, atmospheric, and oceanic sciences; botany and zoology; environmental science; and related fields. As in most matters, only case-by-case analysis can tease out such judgments.

Stepping back a few paces farther to look at the overall systems of science, the balance of benefits looks different yet again. This is because, in short, the Soviet Union was a closed society, while the United States was relatively very open. The shot heard around the world fired by Sputnik in October 1957 was a clear warning that the West ignored what was going on in Soviet science and technology at its own peril. Developments in US and world science beyond the borders of the "Iron Curtain" could be followed through the open, published scientific literature, although with difficulty, since even foreign scientific journals were often kept under lock and key by the Soviet "security organs." Western scientists, however, had no access to developments in Soviet (and East European) science without contact with the people who carried it out, and preferentially by actually going there and spending time there. What they found often surprised them and led to long-lasting and productive scientific collaborations. For all the effort that the West put into building barriers to undesirable "technology transfer" through export controls and other means, nothing they did could come close to the Soviet regime's obsession with secrecy and security. And well they shouldn't have, because the two social systems were rooted in fundamentally opposite models: open societies and closed societies.

Was this to the West's advantage or to the Soviets'? This was a debate that went on not only in public, but also within the US intelligence community itself.[10] It can also be supposed that intelligence agencies in both countries obtained much valuable information, and no doubt much worthless information, from both visitors and hosts after visits had taken place. My guess is actually that the balance of worthless information fell more heavily on the Soviet side, since much of it was likely a rehash of what was available in the open scientific literature but previously unknown in the Soviet Union because even foreign scientific journals were kept under wraps.

This debate, between those who would go to all extents to protect "our secrets" and those who would advocate for opportunities to get access to "their secrets" through cooperative science programs, represented a fundamental tension within the US government not only in the Cold War period but also until at least 1991, if not beyond. My strong impression, as a civilian science manager in an agency that instructed us to keep at arm's length from these discussions, was that the "dark forces" arrayed against scientific cooperation were dominant in the FBI, some of the military research branches, and, depending on the administration, those parts of the Commerce and State Departments that were responsible for export and munitions controls. On the other side—generally, but again with significant variation—was the intelligence community, that part of the State Department responsible for diplomacy, and various civilian mission agencies. At first it seemed very counterintuitive to me that the intelligence community could be the ally of relatively unfettered scientific cooperation, but it proved to be a valuable, though at arm's length, confluence of objectives—from which, I believe, the nation benefited.

Returning to the Cold War period itself, it is probably fair to say that the general public perception of the balance of cooperation with the Soviet Union was that the United States gained more in the cultural and scholarly exchanges while the Soviets and their heirs gained more in the fields of science and technology. Scholars and artists, the reasoning went, could help open up the closed Soviet system while Soviet scientists and engineers were struggling to "catch up" with the West by means of copying, reverse engineering, and outright theft. Though both generalizations have merit, they are by no means accurate in every detail, especially in the science and technology sphere, as the 1977 National Academy of Sciences (NAS) Kaysen Panel report observed. In addition, the Kaysen report pointed to the political and cultural benefits of the exchanges as a less tangible benefit. Loren Graham, one of the report's principal authors, summarizing and commenting on it in *Science* in 1978, argued that, in fact, such benefits were evident: "These links do have political effects that are desirable from the standpoint of the American scientific community. To give one example: 20 years ago if a mathematics researcher at the Steklov Institute in Moscow were arrested, Americans might learn about it 6 months later, a year later, possibly never. Today . . . we will learn about it within several days, and the chances are high that someone in the United States will know that individual personally."[11]

In the public sphere, the balance of benefits issue was a major political football. Members of Congress did not have such nuanced views about who was getting the most from scientific cooperation, and the general public whom they represented had an even less clear picture. The general wisdom was that the Soviets were the crafty and devious winners, and we were the naive and clueless losers. And the irony of this situation was that if a Soviet or Russian historian were to write a parallel history of US-Soviet science cooperation during this and succeeding periods, she would surely depict the mirror image of the same debate in the USSR. No country has a monopoly on either pride or paranoia.

So what's the answer? Who won? The conventional answer, provided by the great historian of US cultural exchange programs, Yale Richmond, is that while the balance of benefits in scientific and technical cooperation favored the Soviets, this deficit was outweighed by the benefits of cultural and scholarly exchanges, as well as the general exposure of Soviet citizens to life and values in America. I shared that view for several years. However, I eventually came to a somewhat different conclusion, that the generally perceived deficit in scientific and technical cooperation, while perhaps not erased much less reversed, was much less enormous than it was thought. I would venture the following general proposition: when an open society interacts with a closed society, the open society wins.

Cooperation—or Assistance? Or, What Went Wrong?

After the fall of the Soviet Union in 1991, as described in chapter 4, the entire calculus of balance of benefits was turned on its head. From a mighty, aggressive global threat and strategic (and moral) rival, the former Soviet Union (FSU) almost overnight became a region whose political collapse and economic weakness became a different and arguably more serious kind of security threat and the source of a feared flood of scientists to the West ("brain drain"). Worries about "loose nukes" and especially unsecured fissile nuclear material achieved nightmare proportions, for good reasons. And whereas prior to 1991 the West would have welcomed with open arms highly qualified people whom it often saw as "defectors" from the Soviet, afterward there were visions of an inundation of scientists that would transform the face of Western laboratories and universities and pose a serious threat to the livelihoods of those countries' scientists. The profusion of programs that followed reflected these new realities and perceptions. The era of "assistance," unthinkable previously, had begun.

Assistance took many forms in the science sphere and those forms themselves evolved over time. The first two embodiments were, in short, bailouts. George Soros's short-lived International Science Foundation (ISF) distributed small sums of money on a competitive merit basis to a very large number of FSU scientists. The stated goal was simply to allow the best ones to continue to work in science, as opposed to abandoning science for other pursuits, from guarding warehouses to banking. The International Science and Technology Center (ISTC) made very large grants to teams of scientists in the closed cities and institutes. The explicit mission here was to "redirect" their work into civilian-oriented research, but it was understood by all that in the short term the most important thing was to provide them with enough income that they would not be tempted to sell their services to rogue states or terrorist groups. In the case of the ISF, and probably for the ISTC as well, a significant portion of the larger grants was used to replace or upgrade obsolete or dysfunctional research instrumentation; a number I frequently heard in the ISF's Long-Term Grants Program was that about 50 percent of the funds went to that purpose. Over the following years, that benchmark held steady for other grant programs, such as CRDF Global.

During these first few years after the Soviet Union's collapse, support coming from these programs accounted for a sizeable portion of the entire domestic research budget in these countries. According to Loren Graham and Irina Dezhina, the support provided by ISF alone in 1994 and 1995 in Russia amounted to approximately 13 percent of Russia's entire expenditures on basic science during that period; while over the entire period from 1993 to 2008, they estimate that "just three organizations—the ISF, the ISTC, and INTAS—have put well over a billion U.S. dollars into Russian science and technology."[12] Surely the ISF's contributions to the research expenditures of other countries in the region was even higher. Comparable figures for the ISTC's impact are not available to my knowledge because military research budgets were not published. But it cannot be doubted that during this period, a very substantial portion of scientific research in the FSU, perhaps in the range of one-quarter to one-third, was funded by foreign sources.

After these early crisis years had passed, it became clear that there had to be some tangible benefit to, and participation from, the United States. Continued unilateral support to former Soviet scientists was not going to be a formula for sustainable funding of these programs, especially with the phase-out of private funding from Soros. Soros believed that the funding of science was primarily a government responsibility, and indeed, it was governments on both sides that needed to be persuaded that cooperative programs had some tangible benefit to both sides.

In the case of the defense-oriented nonproliferation programs, it was certainly enough at first to say that the benefit to the United States—and the world—was global security. In the nonproliferation programs dedicated to physical security (which I do not cover here),[13] the impact of showing that locks were being put on doors and that guards were being paid was well understood. However, with scientists, the US mission agencies were faced with the challenge of proving a negative—that paying former weapons of mass destruction (WMD) scientists to engage in civilian-oriented research was preventing bad things from happening. They used other metrics, such as the number of WMD scientists involved in the programs, the number of closed institutes and laboratories, the number of sustainable civilian jobs created, and even in some cases the wealth created in both countries from the project-related commercial ventures that ensued.

Programs that got underway after the initial crisis period, such as CRDF, sought to build in mutuality of benefit from the outset. CRDF's competitive grant programs funded joint research projects between US and FSU scientists and even between US entrepreneurs and FSU scientists. Mutuality of benefit had to be demonstrated in the proposal and documented in project reports. From the standpoint of funding, the flow of benefits was not quite as balanced, because most of the funds flowed to the FSU side, in terms of individual financial support, travel costs, and, very importantly, research equipment and instrumentation. US participants were awarded funds for project-related travel, but their other costs, including salary, had to come from other sources, such as ongoing standard research grants from agencies such as the National Science Foundation or the National Institutes of Health. This made sense, because it helped to ensure the quality of the projects and of the US participants by presuming, if not requiring, that their core research support came from competitive, peer-reviewed grants awarded by US funders on the basis of scientific merit.

All this being said, I would contend that all of the post-1991 programs I have discussed in this section were first and foremost "assistance" programs. George Soros's ISF was an emergency support program during a time of true economic collapse. The ISTC was also, at the outset, intended to avert a global security emergency of desperate former Soviet WMD scientists looking for ways to employ their talents in exchange for money from nefarious and all-too-willing buyers.

After the immediate economic crisis passed, however, the real issues revolved not so much around starving, desperate scientists as around longer-term issues: Should we help rebuild and upgrade the scientific infrastructure of Russia and the other countries of the FSU? In what ways was that in the national interest of the United States? Was the joint advancement of knowledge a key criterion, and if not, what was?

The 1992 NAS report to science advisor D. Allan Bromley held a clue for some: a healthy science community was essential to ensuring the future of democracy and market economy in the FSU. But there is not universal agreement on this point. From the standpoint of the history of science, Loren Graham has raised troubling questions about whether science and democracy really go hand in hand.[14] Nor is it clear that the entire project of transforming post-Soviet Russia and other countries to Western, market-style democracy, no matter how well-intentioned, was a very good idea to begin with; in this, I agree with Stephen F. Cohen that this effort was not only a failure but that it backfired disastrously.[15] What, then, was the point?

A suggestion might be found in the CRDF's BRHE program. Though it had heavy participation from the United States in terms of oversight and implementation, the sole beneficiaries of the funds were Russians and Russian higher educational institutions. BRHE included no bilateral university-to-university programs, though many of these existed under other auspices. Ultimately, this program's approach of upgrading regional universities throughout Russia and focusing, incidentally, on building universities rather than on the much more scientifically illustrious Russian Academy of Sciences was adopted with a vengeance by the Russian government and is today at the core of its science and education policy. The BRHE program's results were thus positive and tangible. (I must acknowledge here that this program was my personal pride and joy.) But where was the mutual benefit? Or was it just another instance of assistance or noblesse oblige?

I debated this question often with my fellow Americans who helped to design and oversee the BRHE program. It was not project-based scientific research that resulted in advancement of knowledge. It was not in any direct sense the promotion of democracy, though in a sense one could say that scientific peer review is more "democratic" than the command-system allocation of resources from above that was typical of science in the Soviet era. It was not promotion of market economy, though the program did seek to develop a sense of the role of innovation in university research through "technology transfer centers." My personal conclusion was that the benefit to the United States—and the world—was that by reuniting education and research, Russia might develop more open scientific institutions that might help to ensure that its scientists would always be an integral part of the world scientific community and to make it less likely that Russian science would again be hidden from the world as it was when the exchange and cooperative programs discussed in this book made their debut after the Second World War.

The jury is still out on that point. More recently, there is evidence that while the Russian government was happy to adopt the idea of creating "national research universities" as its own, it fiercely denies that the Americans, in particular, had anything to do with it and in fact consider the interest of American foundations in assisting Russian universities as gross interference in its internal affairs. In a scurrilous article in 2016 about The US-Russia Foundation, which sought to build on the BRHE's efforts by enhancing technology transfer at Russian regional universities and offering fellowships in the United States for promising young Russian scientists, Nikolay Sevostyanov wrote that the goal of the foundation's work in this area "is the outright intrusion into Russian universities.... In this regard, an important indication of those intentions is all sorts of lectures which basically consist of propaganda of the political opposition on various levels."[16] It is very likely that the firing of an American venture capitalist, Kendrick White, who had been serving as vice rector of Nizhniy Novgorod State University (NNGU) to advise on technology commercialization, was one of the fruits of this kind of thinking.[17] NNGU was the highly successful pilot project and star of the BRHE program in the late 1990s and early 2000s.

This is just one example of how the dominance of the assistance mind-set among US foundations and policymakers alike proved fatal to even the best of the post-1991 programs. The nonproliferation programs were perhaps the most painful examples of this problem. They were unable, even after twenty years, to rid themselves of the notion that the nonproliferation problem was a Russian problem and to understand that the practices that had been devised to address the issue when assistance arguably was a more appropriate concept were outdated. Even the framers of the Nunn-Lugar Program, Senators Sam Nunn and Richard Lugar, came to understand that the assistance mode was no longer appropriate, but apparently much too late.[18] The Russian government's diplomatic note of July 13, 2011, informing the ISTC of its intent to withdraw from the center by mid-2015 made the handwriting on the wall crystal-clear for all who would see.[19] Yet even in December 2014, when the Nunn-Lugar Program were formally terminated by mutual agreement of the two governments, I would hear brave statements from the US side that it was really only because their mission had been accomplished. We will believe what we want to believe, come what may—or at least we will say so for the public's benefit.

Was there really an opportunity to change course and adopt governing concepts for the nonproliferation programs more appropriate for peer-to-peer cooperation rather than one-sided assistance? I believe the answer is a resounding yes; there were other models and experiences that had been in place for some time, for example the CRDF's cooperative research programs. Tragically, however, the nonproliferation programs became so preoccupied with reacting to congressional criticism that all they could do in terms of policy in those final years was to answer congressional questions and tweak their programs in response rather than undertake a systematic restructuring that would have required congressional trust and commitment that was simply not there.

Perhaps no program model or concept could have saved these and other cooperative programs. The nationalistic fervor, assertiveness, and xenophobia that has become the hallmark of the Putin era may well have ended up hostile to any such effort. But in the end, one-sided efforts based on the assistance model, even the best-designed and best-intended ones, also turned out to be convenient fodder for those who searched for evidence to lay out to the Russian public that these fears were somehow based in reality.

How Well Did the Programs Meet Their Objectives?

It is extremely difficult, if not impossible, to make any conclusive, empirically based judgments about the extent to which the performance of the various bilateral cooperative science programs between the United States and the FSU met their goals. The goals themselves were very complex, and even where quantitative analysis might help—for example, in analyzing bibliometric data—the data are very skimpy due to the Soviets' habit of not publishing in the international scientific literature.

Despite these difficulties, I will venture some conclusions of my own based on the testimony of my interview participants and my own somewhat biased experience of working primarily in the area of basic scientific research cooperation.

The earliest and most consistent goal of scientific cooperation throughout the seventy-year period covered in this book was that of people-to-people diplomacy articulated by President Eisenhower. It was the consensus of my discussants that this goal was a resounding success. It was the personal ties and relationships that were the most enduring—the understanding, reinforced by professional respect, that the partners in the other country were real people, human beings just like everyone else. This is the basis of all mutual understanding; without it, civilization cannot survive.

I would have to rate emergency assistance to top former Soviet scientists in the immediate aftermath of the fall of communism in 1991 to be the second-most-successful outcome. Not only did these various programs—ISF, the Science Centers, CRDF, and others—enable many scientists to simply remain in science, but they also provided lasting tools to help them find support in other ways. The introduction of merit review, and the grantsmanship skills learned through these programs, enabled many scientists in the post-Soviet period to compete successfully for financial support through various programs, especially those of the European Union.

Next would have to come, also in the post-Soviet period, work to prevent the proliferation of materials and technologies related to WMDs. The mutual confidence developed through joint work on purely scientific projects enabled them to work effectively on securing and removing tons of nuclear weapons material, destroying biological weapons facilities, and applying their skills to peaceful, civilian research. It is not clear, however, how lasting the redirection of scientists to civilian research turned out to be, as Russian government funding for defense research has ramped up under President Vladimir Putin, and his government has clamped down on outside access to the closed institutes, which was such a critical component of the nonproliferation programs.

What about the advancement of knowledge itself? It is obvious from the success stories narrated firsthand in this book that these and many other participants achieved significant breakthroughs in scientific knowledge, for example in gravitational physics, paleoclimate, mathematics, nonlinear dynamics, and other fields. This conclusion, however, is tempered by two factors. First, the absence of reliable long-term scientometric data based on publication in the international literature precludes a reliable basis for any real evidence-based analysis; and second, the fact that a great deal of outstanding scientific joint research took place outside the scope of the formal bilateral programs, and the often-disproportionate amount of money spent on protocol events in them, somewhat dilutes the significance of those programs' scientific impact.

Lower in the list, but not insignificant, were outcomes in the very broad area of institutional change, to the extent that this was an explicit goal. As noted above, the introduction of merit review of investigator-initiated, project-based proposals was largely a success with lasting impact, albeit with "variations" to adopt it to local practices. The BRHE's impact on Russian policy regarding increased support to scientific research in universities, with a focus on reuniting research and education, was a particularly impressive outcome, although one loudly disavowed nowadays by loyalists to the Putin government.

The foreign-policy impact of bilateral US-FSU cooperative science programs is quite dubious, in my view. There is no evidence of which I am aware that the programs' existence, or even their periodic suspension or termination, had the slightest impact on Soviet and Russian foreign policy behavior in the big picture—for example, with regard to Afghanistan, Poland, or Ukraine. The habit of US diplomats to use the formal, intergovernmental bilateral science programs as levers to "send messages" to the Soviets and Russians was, in my view, ineffective and meaningless. At the same time, there is reason to believe that negotiations regarding continuation of the private interacademy program in the late 1980s did result in the end of the Soviet government's persecution of Andrey Sakharov. This, however, was not really a foreign policy goal or achievement, but a moral victory on the part of the international scientific community. As far as Henry Kissinger's publicly articulated goal of "and incentive for restraint," disingenuous or not, the public programs had vanishing to zero impact.

Finally, at the very bottom of the list belongs the goal of promoting democracy through instruction and assistance. This was an utter failure. Any such aspirations were purely an American myth. To be sure, the science programs did not concern themselves directly with core democracy issues such as rule of law, elections, political parties, separation of powers, the judiciary, and the like. And they did introduce some rule-based innovations in science administration, such as competitive merit review.

Yet the overall thrust of the US government-sponsored programs of the 1990s and early 2000s was predicated on the notion, as Cohen contends, that we could remake Russia and other post-Soviet countries in our image. This fallacious pursuit, together with the unjustifiable persistence of the noblesse oblige mindset of providing assistance as opposed to peer cooperation, which did permeate the science sphere as well, created deep resentment and ultimately rejection by the Russian government. As George Soros argued in vain in the early 1990s,[20] to effect real change in Russia it would be necessary to have a massive, global Marshall Plan for the region, supported by the world's governments. This was not to be.

Lessons Learned: Eleven Theses

Ultimately, what is more important than how we evaluate what happened in the past is how we benefit from that experience in the future. It will be very difficult to replicate the exact conditions that surrounded US-Soviet and post-Soviet scientific cooperation over the sixty-year period covered here. There are simply not that many countries in the world with an extremely high level of scientific achievement, an enormous community of scientists and engineers, a highly authoritarian political system obsessed with secrecy (for the sake of discussion, by this I refer to the USSR), and the world's largest or second-largest nuclear arsenal. Moreover, where there is free movement of people back and forth and elastic resources, formal programs for international scientific cooperation are often not needed at all.

However, as long as humans seek to advance knowledge and solve global problems, and as long as there are geopolitical divisions and confrontations, there will always be a desire and a need for scientists to reach out beyond national borders and work with other scientists. In that spirit, here are eleven theses in which I have sought to encapsulate what there is to learn from the experience covered in this book:

I.

What is most important about international scientific collaboration is not that it is international but that it is collaboration.

II.

Science not only knows borders but also is enriched by them. Borders of culture, language, geography, nations, political systems, and disciplines can be both nuisances and opportunities. It all depends on what you do with them.

III.

In terms of the advancement of knowledge, international scientific collaboration is more effective when its financial support is based on its scientific merit rather than its international impact.

IV.

The value of specially organized and funded programs to promote international scientific cooperation lies not in the amount of money they provide, but in the opportunities they create to engage in cross-border collaboration that may not be available through traditional means of support.

V.

The most effective and most enduring programs of international scientific cooperation are those that are jointly conceived, jointly designed, jointly governed, jointly funded, and jointly implemented.

VI.

Programs to provide assistance to foreign scientific communities can be extremely effective in the short-term if they are jointly conceived, designed, governed, funded, and implemented and if they have broad institutional reach. However, indefinite perpetuation of the assistance rationale can create resentment and hostility in the host country, causing those programs to fail and even damage the bilateral relationship.

VII.

The role of private foundations and nongovernmental organizations in promoting international scientific cooperation cannot be overestimated. These organizations can be nimbler and more effective than government agencies of any kind in identifying and framing opportunities and approaches in countries and global regions where political and other conditions are not ideal for free and open scientific communication and logistical support.

VIII.

The private, for-profit sector can be a major actor in international scientific cooperation, if facilitating programs are sufficiently creative and well-conceived. When private companies generate financial profit by creating value and solving problems in the global knowledge economy, they can be allies in international scientific cooperation.

IX.

"The political significance of international science and technology agreements ends the moment they are signed."[21] This is to say that the activities that take place under such agreements must be demonstrably meritorious in terms of their science, and if so, they are likely to continue even beyond the agreement's lifetime. Conversely, if the responsible mission agencies do not provide the resources to implement them in support of their mission, the agreements will fail to achieve any real purpose.

X.

When governments pursue formal bilateral international scientific cooperative programs as an instrument of foreign policy, they must be prepared to use that instrument as a "stick" when extreme circumstances warrant it; otherwise, the programs' meaning as foreign policy instruments is lost, and they will stand to lose public support. However, the overuse of such programs to "send messages" in less extreme circumstances is highly ineffective and can do more harm than good to the scientists in the offending country who take part in them.

XI.

The point is not only to understand the world but also to change it.[22]

Notes

1. In a speech to an international conference at the Royal Society in London in October 2001, cosponsored by the Royal Society and CRDF, on ten years of scientific cooperation with the FSU.
2. Interview with Loren Graham, October 19, 2015.
3. The sister of the late physicist Richard Feynman.
4. Joan Feynman, in interview with Alexander Ruzmaikin and Joan Feynman, October 6, 2015.
5. See the discussion in chapter 5, "The Science," of the heated exchange between Executive Director of AAAS William D. Carey and Deputy Secretary of Defense Frank Carlucci.
6. See the section in chapter 9 titled, "Intelligence Gathering and Secrecy."
7. *Review of U.S.-USSR Interacademy Exchanges and Relations* 1977, 90–92.
8. Graham 1972.
9. Ibid, pp. 257–96.
10. See McGrath (1962) for a very revealing internal discussion within the CIA about this issue.
11. Graham 1978, 387.
12. Graham and Dezhina 2008, 95, 89.
13. For example, the Material Protection, Control, and Accounting (MPC&A) program of the National Nuclear Security Administration.
14. Graham 1998.
15. Cohen 2000.
16. Sevostyanov 2016.
17. Sinelschikova 2015.
18. See Nunn and Lugar 2015.
19. See Schweitzer 2013, 263.
20. See Soros 2000, and chapter 4, "Major New Programs Come—and Go."
21. A maxim attributed to Herman Pollack, the State Department's first director of its bureau of scientific and technological affairs, as related to me by Arthur E. Pardee Jr.
22. A gloss on Marx's dictum "The philosophers have only interpreted the world, in various ways; the point, however, is to change it." Karl Marx, "Theses on Feuerbach," in Tucker 1972, 109.

APPENDIX

List of Interviews

Maia Alkalkatsi
　Tbilisi, Georgia
　December 7, 2015
　Institute of Botany, Ilia State
　University

Viktor Baryakhtar
　Kyiv, Ukraine
　December 10, 2015
　Vice President, National Academy
　of Sciences of Ukraine; Institute of
　Magnetism, National Academy of
　Sciences of Ukraine

David Bell
　Minneapolis, Minnesota
　January 14, 2016
　President, Phygen Coatings Inc.;
　Board Member, United States
　Industry Coalition

Julie Brigham-Grette
　Washington, DC
　December 2, 2015
　Department of
　Geosciences, University of
　Massachusetts–Amherst

Cathleen Campbell
　Arlington, Virginia
　September 16, 2015
　President and CEO, CRDF Global*

Rita Colwell
　College Park, Maryland
　June 9, 2016
　Distinguished University Professor,
　University of Maryland at College
　Park; Director, National Science
　Foundation*

Lawrence Crum
　Seattle, Washington
　January 6, 2016
　Applied Physics Laboratory,
　University of Washington

Irina Dezhina
　Washington, DC
　November 3, 2015
　Head, Research Group on Science
　and Industrial Policy, Skolkovo
　Institute of Science and
　Technology

Cassandra Dudka
　Arlington, Virginia
　July 30, 2015
　Office of International Science
　and Engineering, National Science
　Foundation

Vladimir Elisashvili
　Tbilisi, Georgia
　December 7, 2015
　Director, Animal Husbandry
　and Feed Production Institute,
　Agrarian University of
　Georgia

Oleg Federov
　Kyiv, Ukraine
　December 16, 2015
　Director, Institute of Space
　Research, National Academy of
　Sciences of Ukraine

(Continued)

Joan Feynman
    Pasadena, California
    October 6, 2015
    Jet Propulsion Laboratory*

Margaret Finarelli
    Arlington, Virginia
    August 4, 2016
    Associate Administrator for Space Policy and International Relations*

Maxim Frank-Kamenetskii
    Boston, Massachusetts
    Ocotber 19, 2015
    Department of Biomedical Engineering, Boston University

Yuriy Gorobets
    Kyiv, Ukraine
    December 10, 2015
    Director, Institute of Magnetism, National Academy of Sciences of Ukraine; Minister of Education, Ukraine*

Rose Gottemoeller
    Washington, DC
    April 21, 2016
    Undersecretary of State for Arms Control and International Security, Department of State

Loren Graham
    Boston, Massachusetts
    October 19, 2015
    History of Science Program, Massachusetts Institute of Technology*

Randolph Guschl
    Wilmington, Delaware
    March 21, 2016
    Director of Global Research and Development, DuPont Corporation*

F. Gray Handley
    Washington, DC
    April 20, 2016
    Associate Director for International Affairs, National Institute of Allergy and Infectious Diseases, National Institutes of Health

Paul Hearn
    Washington, DC
    August 5, 2015
    Senior Scientist, US Geological Survey*

Siegfried Hecker
    Palo Alto, California
    January 28, 2016
    Director, Los Alamos Scientific Laboratory*

Roald Hoffmann**
    Boston, Massachusetts
    February 18, 2016
    Department of Chemistry, Cornell University

Laura Holgate***
    Washington, DC
    October 13, 2015
    Senior Director, Weapons of Mass Destruction, Terrorism and Threat Reduction, National Security Council*, US Ambassador to the IAEA*

Zurab Javakishvili
    Tbilisi, Georgia
    December 8, 2015
    Director, Institute of Earth Sciences, Ilia State University

Moisei Kaganov
    Boston, Massachusetts
    October 16, 2015
    Institute of Physical Problems, USSR Academy of Sciences*

George Kamkamidze
  Tbilisi, Georgia
  December 7, 2015
  Director, "Neolab" Clinic, Medical State University

Ramaz Katsarava
  Tbilisi, Georgia
  December 8, 2015
  Director of Chemistry and Molecular Engineering, Agrarian University of Georgia

Barbara Kilosandidze
  Tbilisi, Georgia
  December 8, 2015
  Head of Laboratory for Holography Recording and Information Processing, Institute of Cybernetics

John Kiser
  Sperryville, Virginia
  March 30, 2016
  Author; President, Kiser Research Inc.*

Elizabeth Kutter
  Olympia, Washington
  January 5, 2016
  Professor of Biology, Evergreen State College

John Logsdon
  Washington, DC
  November 17, 2015
  George Washington University

Terry Lowe
  Golden, Colorado
  March 26, 2016
  Department of Metallurgy, Colorado School of Mines; Head, Department of Metallurgy, Los Alamos Scientific Laboratory*

Igor Lytvynov
  Kyiv, Ukraine
  December 16, 2015
  Deputy Director, Science and Technology Center, Ukraine

Eileen Malloy***
  Seattle, Washington
  January 6, 2016
  US Ambassador to Kazakhstan;* US Embassy Moscow*

Boris Movchan
  Kyiv, Ukraine
  December 17, 2015
  Director, International Center for Electron Beam Technologies

Anton Naumovets
  Kyiv, Ukraine
  December 15, 2015
  Vice President, National Academy of Sciences of Ukraine; Institute of Magnetism, National Academy of Sciences of Ukraine

Norman Neureiter
  Arlington, Virginia
  January 12, 2016
  Director, Program on Science, Technology and Public Policy, AAAS; Science Advisor to the Secretary of State*

Thomas Pickering***
  Washington, DC
  September 24, 2015
  Undersecretary of State for Political Affairs*; US Ambassador to the United Nations*; US Ambassador to Russia*; Vice President for International Affairs, Boeing*

Marilyn Pifer
  Arlington, Virginia
  October 27, 2015

(Continued)

Director, Capacity Building, CRDF
Global

Peter Raven
St. Louis, Missouri
February 23, 2016
Chairman, Exploration Committee,
National Geographic Society;
Director, Missouri Botanical
Gardens*; Board Chair, CRDF*

Alexander Ruzmaikin
Pasadena, California
October 6, 2015
Jet Propulsion Laboratory; Institute
of Applied Physics, USSR Academy
of Sciences*

Sergiy Ryabchenko
Kyiv, Ukraine
December 14, 2015
Institute of Physics, National
Academy of Sciences of Ukraine;
Minister of Education and Science,
Ukraine*

Ekaterine Sanaia
Tbilisi, Georgia
December 8, 2015
Deputy Director, Sokhumi
Institute of Physics and
Technology

Glenn Schweitzer
Washington, DC
February 26, 2016
Director, Office for Central Europe
and Eurasia, National Academy of
Sciences

Marjorie Senechal
Stockbridge, Massachusetts
May 10, 2016
Department of Mathematics, Smith
College*
Co-Chair, Basic Research and
Higher Education in Russia
program*

Boris Shklovskii
Minneapolis, Minnesota
January 14, 2016
University of Minnesota
Ioffe Physico-Technical Institute,
Academy of Sciences of the USSR,
St. Petersburg, Russia*

Revaz Solomonia
Tbilisi, Georgia
December 8, 2015
Director, Institute of Chemical
Biology, Ilia State University

Valery Soyfer
Washington, DC
October 29, 2015
School of Systems Biology, George
Mason University
Director, Soros International
Science Education Program*

Marina Tediashvili
Tbilisi, Georgia
December 8, 2015
Head, Microbiology Laboratory,
Eliava Institute of Bacteriophage,
Microorganizsms and
Bacteriophages

James Thompson
Columbia, Missouri
February 24, 2016
College of Engineering, University
of Missouri; Board Chair, United
States Industry Coalition

Kip Thorne**
Pasadena, California
October 6, 2015
Department of Physics, California
Institute of Technology

Tengiz Tsertsvadze
Tbilisi, Georgia
December 7, 2015
Board Chairman, Infectious
Disease, AIDS and Clinical

Immunology Research Center;
National AIDS Coordinator

Alexander Vilenkin
　Medord, Massachusetts
　October 15, 2015
　Director, Institute of Cosmology,
　Tufts University

Carol Vipperman
　Seattle, Washington
　January 8, 2016
　President and CEO, Foundation
　for Russian-American Economic
　Cooperation*

Gary Waxmonsky
　Washington, DC
　October 9, 2015
　Chief of Staff, Office of
　International Affairs,
　Environmental Protection Agency*

Andrew Weber
　Arlington, Virginia
　April 25, 2016
　Assistant Secretary of Defense for
　Nuclear, Chemical, and Biological
　Defense Programs*

Karl Western
　Rockville, Maryland
　July 15, 2016
　Senior International Scientific
　Advisor, National Institute of
　Allergy and Infectious Diseases,
　National Institutes of Health

Deborah Wince-Smith
　Washington, DC
　July 11, 2016
　President and CEO, Council on
　Competitiveness

Yaroslav Yatskiv
　Kyiv, Ukraine
　December 15, 2015
　Director, Main Astronomical
　Observatory, National Academy of
　Sciences of Ukraine;* Presidium,
　National Academy of Sciences of
　Ukraine

Konstantin Yushchenko
　Kyiv, Ukraine
　December 15, 2015
　Deputy Director, E. O. Paton
　Electric Welding Institute

Mikhail Zgurovsky
　Kyiv, Ukraine
　December 14, 2015
　Rector, Kiev Polytechnic Institute

John Zimmerman
　Washington, DC
　September 30, 2015
　Science Counselor,
　US Embassy,
　Moscow (twice)*;
　Soviet Desk,
　Department of State*;
　Office of Naval Research*

* Former position (at time of interview)
** Nobel Laureate
*** Ambassador

BIBLIOGRAPHY

"About Pugwash." Pugwash Conferences on Science and World Affairs. Accessed June 12, 2015. http://pugwash.org/about-pugwash/.

Ailes, Catherine F., and Arthur E. Pardee Jr. *Cooperation in Science and Technology: An Evaluation of the U.S.-Soviet Agreement.* Boulder, CO: Westview Press, 1986.

Albrecht, Mark. *Falling Back to Earth: A First Hand Account of the Great Space Race and the End of the Cold War.* New Media Books, 2011.

Alfimov, Mikhail V., Aleksandr F. Andreev, E. P. Velikhov, Yurii M. Kagan, N. N. Kudryavstev, V. A. Matveev, Oleg V. Rudenko, Aleksandr Yu Rumyantsev, Roald Z. Sagdeev, Vladimir E. Fortov, Aleksei M. Fridman, and V. E. Cherkovets. "In Memory of Aleksandr Mikhailovich Dykhne." *Physics-Uspekhi* 48, no. 2 (2005).

Ambegoakar, V. "The Landau School and the American Institute of Physics translation program." *Uspekhi Fizicheskikh Nauk* 51, no. 12 (2008): 1287–90. This article was kindly made available to me by the editorship of *Uspekhi*.

AmeRus Foundation for Research and Development Act of 1992, HR 4550, 102nd Congress. Accessed August 7, 2015. http://thomas.loc.gov/cgi-bin/query/z?c102:H.R.4550.IH.

Aronova, Elena. "Big Science and 'Big Science Studies' in the United States and the Soviet Union during the Cold War." In *Science and Technology and the Global Cold War*, edited by Naomi Oreskes and John Krige, 393–429. Cambridge, MA: MIT Press, 2014.

"The Atomic Scientists of Chicago." *Bulletin of the Atomic Scientists* 1, no. 1 (December 10, 1945): 1. Accessed June 18, 2015. https://books.google.ca/books?id=-wsAAAAAMBAJ.

Balzer, Harley D. "Is Less More? Soviet Science in the Gorbachev Era." *Issues in Science and Technology* 1, no. 4 (Summer 1985): 29–46.

———. "Russia's Scientists Fall Silent." *New York Times*, July 22, 2015. Accessed May 18, 2017. https://www.nytimes.com/2015/07/23/opinion/russias-scientists-fall-silent.html.

Balzer, Harley D., and Stephen Sternheimer. *Soviet Science on the Edge of Reform.* Boulder, CO: Westview Press, 1989.

"Basic Science Research and Higher Education in Russia: A Proposal for Reform." Internal CRDF report, August 2007.

Beissinger, Mark R. *Scientific Management, Socialist Discipline, and Soviet Power.* Cambridge, MA: Harvard University Press, 1988.

Bergelsdorf, Irving S. "Letter to Pravda." *Science* 182, no. 4110 (October 26, 1973): 334.

Bhandari, Rajika, and Raisa Belyavina. "Evaluating and Measuring the Impact of Citizen Diplomacy: Current Status and Future Directions." New York, Institute of International Education, 2011. Accessed September 22, 2015. https://www.iie.org/Research-and-Insights/Publications/Evaluating-Measuring-Impact-of-Citizen-Diplomacy.

"Bilateral Co-funding Programs: U.S.-Russia Bilateral Collaborative Research Partnerships on Cancer." PowerPoint briefing presented to the National Cancer Institute's Board of Scientific Advisors, March 28, 2016. Accessed 29 July 2016. http://deainfo.nci.nih.gov/advisory/bsa/0316/1300Pearlman.pdf.

The Biological Threat Reduction Program of the Department of Defense: From Foreign Assistance to Sustainable Partnerships. National Research Council Report. Washington, DC: National Academies Press, 2007.

Brands, H. W. *Reagan: The Life.* New York: Anchor, 2015.

A Bridge to Partnership in Research: Activities over the FP6 Period 2002–2006. Brussels: International Association for the Promotion of Co-Operation with Scientists from the New Independent States of the Former Soviet Union, ca. 2006.

Brigham-Grette, J., M. Melles, P. Minyuk, A. Andreev, P. Tarasov, R. DeConto, S. Koenig, et al. "Pliocene Warmth, Polar Amplification, and Stepped Pleistocene Cooling Recorded in NE Arctic Russia." *Science* 340, no. 6139 (June 21, 2013): 1421–27. https://doi.org/10.1126/science.1233137; plus supplement.

Bromberg, Joan Lisa. *Fusion: Science, Politics, and the Invention of a New Energy Source.* Cambridge, MA: MIT Press, 1982.

Brzezinski, Matthew. *Red Moon Rising: Sputnik and the Hidden Rivalries That Ignited the Space Age.* New York: Times Books, 2007.

Byrnes, Robert F. *Soviet-American Academic Exchanges, 1958–1975.* Bloomington: Indiana University Press, 1976.

Carey, William D. "Scientific Exchanges and National Security." Letter to Deputy Secretary of Defense Frank Carlucci. *Science* 215, no. 429 (January 8, 1982): 139–141.

Carter, Ashton B., Kurt Campbell, Steven Miller, and Charles Zraket. *Soviet Nuclear Fission: Control of the Nuclear Arsenal in a Disintegrating Soviet Union.* Cambridge, MA: Belfer Center for Science and International Affairs, Harvard Kennedy School, November 1991.

Chamberlain, Owen. "Scientists Protest Trials in USSR." *Bulletin of the Atomic Scientists* 34, no. 8 (October, 1978): 9, 50–51.

Cohen, Stephen F. *Failed Crusade: America and the Tragedy of Post-Communist Russia.* New York: Norton, 2000.

Coriden, Guy E. "The Intelligence Hand in East-West Exchange Visits." *Studies in Intelligence* 2, no. 3 (1958): 63–70. Accessed July 12, 2015. https://www.cia.gov/library/center-for-the-study-of-intelligence/kent-csi/vol2no3/html/v02i3a09p_0001.htm.

David-Fox, Michael. *Showcasing the Great Experiment: Cultural Diplomacy and Western Visitors to the Soviet Union, 1921–1941.* New York: Oxford University Press, 2012.

Demina, Natalya, and Gennadiy Mesyats. "Sverzheniye Fortova" ["Fortov's Ouster"]. *Troitskiy variant,* no. 225 (March 2017): 1–2. Accessed 17 April 2017. http://trv-science.ru/2017/03/28/sverzhenie-fortova/.

Dezhina, Irina G. *The International Science Foundation: The Preservation of Basic Science in the Former Soviet Union.* Translated from the Russian by Boris Gorney. New York: International Science Foundation, 2000.

———. "Science and Innovation Policy of the Russian Government: A Variety of Instruments with Uncertain Outcomes?" *Public Administration Issues.* Special Issue (electronic edition, 7–26 (in English). https://doi.org/10.17323/1999-5431-2017-0-5-7-26.

———. "'Utechka umov' iz post-Sovietskoy Rossii: Evolutsia yavleniya i ego otsenok" ["The 'Brain Drain' from Post-Soviet Russia: The Phenomenon's Evolution and an Assessment"]. *Naukovedenie,* no. 3 (2002): 25–56.

Eisenhower, Dwight D. "People to People Program." From Dwight D. Eisenhower Presidential Library, Museum and Boyhood Home: Abilene, Kansas. Accessed

July 20, 2015. http://www.eisenhower.archives.gov/research/online_documents/people_to_people.html.

———. "Remarks at the People-to-People Conference. September 11, 1956." American Presidency Project. Accessed June 18, 2015. http://www.presidency.ucsb.edu/ws/?pid=10599.

———. *White House Years: Waging Peace, 1956–1961*. Garden City, NY: Doubleday, 1965.

Eisenhower, Susan. *Partners in Space: US-Russian Cooperation after the Cold War*. Washington, DC: Eisenhower Institute, 2004.

Foreign Operations, Export Financing, and Related Programs Appropriation Act, 1994, Pub. L. No. 103-87, Sec. 575, 107 STAT 972-773 (1993). Accessed August 18, 2015. https://www.congress.gov/bill/103rd-congress/house-bill/2295/text?q=%7B%22search%22%3A%5B%22Pub.+No.+103-87+575%2C+107+Stat.+972-773+%281993%29%22%5D%7D&resultIndex=1.

Fortov, Vladimir Ye. *Osnovniye napravleniya rasvitiya Rossiyskoy akademii nauk: Vyborna programma V. Ye. Fortova* ["Basic Directions of Development of the Russian Academy of Sciences: The Electoral Program of V. Ye. Fortov"]. Moscow, Russia. 2017.

FREEDOM Support Act, Pub. L. No. 102-511 106 Stat. 3320 (1992). Accessed August 16, 2015. http://www.gpo.gov/fdsys/pkg/STATUTE-106/pdf/STATUTE-106-Pg3320.pdf.

"Glava Minobrnauki pristanovila protsess ob'yedinyeniya vuzov," *Polit.ru.* September 26, 2016. Accessed September 27, 2016. http://www.polit.ru/news/2016/09/26/vasilieva.

Global Security Engagement: A New Model for Cooperative Threat Reduction; Report of the National Academy of Sciences. Washington, DC: National Academies Press, 2009.

Gorwitz, Mark. "Vyacheslav Danilenko—Background, Research, and Proliferation Concerns." *Reports of the Institute for Science and International Security*, November 29, 2011. Accessed August 1, 2016. http://isis-online.org/isis-reports/detail/vyacheslav-danilenko-background-research-and-proliferation-concerns/8.

Graham, Loren R. "How Valuable Are Scientific Exchanges with the Soviet Union?" *Science* 202, no. 4366 (October 7, 1978): 383–90.

———. *Lonely Ideas: Can Russia Compete?* Cambridge, MA: MIT Press, 2013.

———. *Lysenko's Ghost: Epigenetics and Russia*. Cambridge, MA: Harvard University Press, 2016.

———. *Science and Philosophy in the Soviet Union*. New York: Alfred Knopf, 1972.

———. *Science in Russia and the Soviet Union: A Short History*. New York: Cambridge University Press, 1993.

———. "Scientists, Human Rights, and the Soviet Union." *Bulletin of the Atomic Scientists* 42, no. 4 (April 1986): 8–10.

———. *The Soviet Academy of Sciences and the Communist Party, 1927–1932*. Princeton, NJ: Princeton University Press, 1977.

———. *What Have We Learned about Science and Technology from the Russian Experience?* Palo Alto, CA: Stanford University Press, 1998.

Graham, Loren R., and Irina D. Dezhina. *Science in the New Russia: Crisis, Aid, Reform*. Bloomington: Indiana University Press, 2008.

Grodzins, Morton, and Eugene Rabinowitch, eds. *The Atomic Age: Scientists in National and World Affairs*. New York: Basic Books, 1963.

Gwertzman, Bernard. "Soviet's [sic] Said to Have Warned U.S. of Retaliation in Arrest of Russians." *New York Times*, June 15, 1978, 1.

Hecker, Siegfried S., ed. *Doomed to Cooperate: How American and Russian Scientists Joined Forces to Avert Some of the Greatest Post–Cold War Nuclear Dangers*. 2 vols. Los Alamos, NM: Bathtub Row Press, 2016.

Hoffman, David E. *The Dead Hand: The Untold Story of the Cold War Arms Race and Its Dangerous Legacy*. New York: Anchor Books, 2009.

"IC-INTAS—International Association for the Promotion of Cooperation with Scientists from the Independent States of the Former Soviet Union (INTAS), 1993-." *Community Research and Development Information Service (CORDIS) of the European Community*. Accessed July 2, 2016. http://cordis.europa.eu/programme/rcn/493_en.html.

"Integration of Teaching and Scientific Research in Russia: An Independent Evaluation of the Basic Research and Higher Education Program 1998–2007." Internal report of CRDF Global, December 2007.

Jamgotch, Nish, Jr., ed. *Sectors of Mutual Benefit in U.S.-Soviet Relations*. Durham, NC: Duke University Press, 1985.

Khrushchev, Nikita S. "Khrushchev's Secret Speech, 'On the Cult of Personality and Its Consequences,' Delivered at the Twentieth Party Congress of the Communist Party of the Soviet Union," February 25, 1956, History and Public Policy Program Digital Archive, From the Congressional Record: Proceedings and Debates of the 84th Congress, 2nd Session (May 22, 1956–June 11, 1956), C11, Part 7 (June 4, 1956): 9389–9403. https://digitalarchive.wilsoncenter.org/document/115995

———. "On Peaceful Coexistence." *Foreign Affairs* 38, no. 1: (October 1959).

Kiser, John W., III. *Communist Entrepreneurs: Unknown Innovators in the Global Economy*. New York: Franklin Watts, 1989.

———. *Report on the Potential for Technology Transfer from the U.S.S.R. to the United States*. Washington, DC: US Department of State, 1977.

———. "Tapping Eastern Bloc Technology." *Harvard Business Review*, March–April 1982: 85–92.

———. "Technology Is Not a One-Way Street." *Foreign Policy*, June 1, 1976, 135–48.

———. "What Gap? Which Gap?" *Foreign Policy*, September 1, 1978, 90–94.

Kutter, Elizabeth. "About Us." *Evergreen Phage Lab* (blog). Accessed March 16, 2019. http://blogs.evergreen.edu/phage/about/.

Lederer, William J., and Eugene Burdick. *The Ugly American*. New York: W.W. Norton & Company, 1958.

Lemothe, Dan. "Russian Surveillance Plane Soars over the Pentagon, Capitol and Other Washington Sites." *Washington Post*, August 9, 2017, A5. Accessed August 10, 2017. https://www.washingtonpost.com/news/checkpoint/wp/2017/08/09/russian-surveillance-plane-soars-over-the-pentagon-capitol-and-other-washington-sights/?hpid=hp_local-news_cp-planeregional-6pm%3Ahomepage%2Fstory&utm_term=.e9619eb07122.

Logsdon, John M., and James R. Millar. "U.S.-Russian Cooperation in Human Space Flight Assessing the Impacts." Unpublished paper of the Space Policy Institute and Institute for European, Russian, and Eurasian Studies, Elliott School of International Affairs, George Washington University, Washington, DC, February 2001. Accessed May 19, 2015. http://www.nasa.gov/externalflash/iss-lessons-learned/docs/partners_us_russia.pdf.

Lubrano, Linda L. "The Political Web of Scientific Cooperation between the U.S.A. and USSR." In Jamgotch 1985, 50–82.

Maggs, Peter. "No Honorable Solution in 1978 Soviet Case of U.S. Businessman." Letter to *New York Times*, September 23, 1986.

Markusova, V. A., A. N. Libkind, L. E. Mindeli, and M. Jansz. "Research Performance by Federal and National Research Universities and Impact of Competitive Funding on Their Publication Activity." *COLLINET Journal of Scientometrics and Information Management* 7, no. 2 (December 2013): 1–13.

Markusova, V. A., A. N. Libkind, M. Jansz, and L. E. Mindeli. "Bibiometric Performance in Two Main Research Domains: The Russian Academy of Sciences and the Higher Education Sector." *COLLINET Journal of Scientometrics and Information Management* 8, no. 1 (June 2014): 1–12.

McConnell, Allen. "The Origin of the Russian Intelligentsia." *Slavic and East European Journal* 8, no. 1 (Spring 1964): 1–16.

McGrath, James. "The Scientific and Cultural Exchange." Central Intelligence Agency, ca. 1962. Historical document declassified in 1994. Accessed July 12, 2015. https://www.cia.gov/library/center-for-the-study-of-intelligence/kent-csi/vol7no1/html/v07i1a02p_0001.htm.

Medvedev, Zhores A. *Rise and Fall of T. D. Lysenko.* New York: Columbia University Press, 1969.

Melles, M., J. Brigham-Grette, P. Minyuk, et al. "2.8 Million Years of Arctic Climate Change from Lake El'gygytgyn, NE Russia." *Science* 337 (2012): 315–20, plus supplement.

Mitchell, Lawrence C. "Soviet-American Exchange of Scientists." *Bulletin of the Atomic Scientists* 18, no. 2 (February 1962): 15–17.

New Frontiers in Science Diplomacy: Navigating the Changing Balance of Power. London: Royal Society, 2010. Accessed September 22, 2015. http://www.aaas.org/sites/default/files/New_Frontiers.pdf.

"New Ministry for Russian Science." *Science* 360, no. 6391 (May 25, 2018): 838.

Nuclear Nonproliferation: DOE's Program to Assist Weapons Scientists in Russia and Other Countries Needs to Be Reassessed. Report GAO-08-189. Washington, DC: US Government Accountability Office, December 2007.

Nunn, Sam, and Richard Lugar. "The United States and Russia Must Repair Their Partnership on Nuclear Security." *Washington Post*, January 23, 2015. Accessed August 29, 2016. https://www.washingtonpost.com/opinions/the-united-states-and-russia-must-repair-their-partnership-on-nuclear-security/2015/01/23/555b9a60-a271-11e4-903f-9f2faf7cd9fe_story.html?utm_term=.0373f2c61d9f.

O'Rourke, P. J. "Why Does the USA Depend on Russian Rockets to Get Us into Space?" *Daily Beast*, June 22, 2014. Accessed July 30, 2017. http://www.thedailybeast.com/why-does-the-usa-depend-on-russian-rockets-to-get-us-into-space.

"Outpost of an Empire." https://www.fortross.org/russian-american-company.htm. Accessed December 14, 2018.

Overcoming Impediments to U.S.-Russia Cooperation on Nuclear Nonproliferation: Report of a Joint Workshop. Washington, DC: National Academies Press, 2004. Accessed May 19, 2015. https://books.google.com/books?id=QyqtxwgbbToC&printsec=frontcover#v=onepage&q&f=false.

Peer Review of the Ukrainian Research and Innovation System. Brussels: European Union, 2016.

Penfield, Wilder. "Science Knows No Boundaries." *Science* 147, no. 3660 (February 19, 1965): 909–10.

"President: Ukrainian Science Should Become Modern, Efficient and Innovative." Official website of the president of Ukraine, Petro Poroshenko, December 25, 2015. Accessed August 13, 2016. http://www.president.gov.ua/en/news/ukrayinska-nauka-maye-stati-suchasnoyu-efektivnoyu-innovacij-36564.

"Pro naukovu I naukovo-tekhnichnu diyal'nist'" ["On Scientific and Scientific-Technical Activity"]. *Vidomosti verkhovnoi Radi (VVR)*, no. 3 (2016): 25.

"Pugwash Meetings 1957–2013: An Overview." Accessed June 18, 2015. https://pugwashconferences.files.wordpress.com/2014/05/20131101_pugwashmtgs_1957_2013_rev2.pdf.

"Putin Deplores Collapse of USSR." *BBC News*, April 25, 2005. Accessed June 25, 2016. http://news.bbc.co.uk/2/hi/4480745.stm.

Rabkin, Yakov M. *Science between the Superpowers*. New York: Twentieth Century Fund, 1988.

Reagan, Ronald. "Address to the Nation and Other Countries on United States-Soviet Relations, January 16, 1984." Accessed August 15, 2016. https://reaganlibrary.archives.gov/archives/speeches/1984/11684a.htm.

———. "Evil Empire Speech." Address to the National Association of Evangelicals, Orlando, Florida, March 8, 1983. Accessed December 12, 2018 at http://voicesofdemocracy.umd.edu/reagan-evil-empire-speech-text/.

Reischauer, Edwin O. "The Broken Dialogue with Japan." *Foreign Affairs* 39, no. 1: (October 1960).

Reorientation of the Research Potential of the Former Soviet Union: A Report to the Assistant to the President for Science and Technology. Washington, DC: National Academy Press, 1992.

Review of the US/USSR Agreement on Cooperation in the Fields of Science and Technology. Washington, DC: National Academy of Sciences, 1977.

Review of U.S.-U.S.S.R. Cooperative Agreements on Science and Technology: Special Oversight Report No. 6. Subcommittee on Domestic and International Scientific Planning and Analysis of the Committee on Science and Technology. US House of Representatives. 94th Congress, second sess. Washington, DC: US Government Printing Office, 1976.

Review of U.S.-USSR Interacademy Exchanges and Relations. Washington, DC: National Academy of Sciences, 1977.

Richmond, Yale. "Cultural Exchange and the Cold War: How the West Won." American Diplomacy: Foreign Service Dispatches and Periodic Reports on US Foreign Policy, March 2013. Accessed June 12, 2015. http://www.unc.edu/depts/diplomat/item/2013/0105/ca/richmond_exchange.html.

———. *Cultural Exchange and the Cold War: Raising the Iron Curtain*. University Park: Pennsylvania University Press, 2003.

———. *U.S.-Soviet Cultural Exchanges, 1958–1986: Who Wins?* Boulder, CO: Westview Press, 1987.

Rojansky, Matthew, and Izabella Tabarovsky. "The Latent Power of Health Cooperation in U.S.-Russia Relations." *Science and Diplomacy*, June 2013.

Roth, Andrew. "The Kremlin Is Done Betting on Trump and Planning How to Strike Back against U.S. Sanctions." *Washington Post*, July 29, 2017. Accessed July 30, 2017. https://www.washingtonpost.com/world/europe/the-kremlin-is-done-betting-on-trump-and-planning-how-to-strike-back-against-us-sanctions/2017/07/29/6ae6dfe4-714e-11e7-8c17-533c52b2f014_story.html?utm_term=.64344c16f03b.

Rubbia, Carlo. Letter to François Mitterand dated September 26, 1991. CERN Archive. I am grateful to CERN for kindly making this important document available to me.

"Russian Roulette." *Nature* 499, no. 5–6: (July 4, 2013). Editorial.

"Russian Science Foundation Shuts Down after Being Branded 'Foreign Agent.'" *The Guardian* (US edition), July 8, 2015. Accessed August 13, 2016. https://www.theguardian.com/world/2015/jul/08/russian-science-dynasty-foundation-branded-foreign-agent-kremlin.

Sagdeev, Roald Z. "Changes in Science: East & West." In *Beyond the Cold War: The Changing Arena of Science*, edited by Judy Jackson, Joseph Lach, and John Venard, 12–31. Roundtable on September 9–10, 1993, sponsored by Fermi National Accelerator

Laboratory and Fermilab Industrial Affiliates. Batavia, IL: Fermi National Accelerator Laboratory, 1993. Accessed July 22, 2016. http://lss.fnal.gov/conf/C9309091/1993.pdf.

———. *The Making of a Soviet Scientist: My Adventures in Nuclear Fusion and Space from Stalin to Star Wars*, edited by Susan Eisenhower. New York: John Wiley, 1994.

———. "Science and Perestroika: A Long Way to Go." *Issues in Science and Technology* 4, no. 4 (Summer 1988): 48–52.

Saltykov, Boris G. "The Reform of Russian Science." *Nature* 388, no. 6637 (July 3, 1997): 16–18.

Schiermeier, Quirin. "Vote Seals Fate of Russian Academy of Sciences." *Nature*, September 19, 2013. Accessed April 17, 2017. http://www.nature.com/news/vote-seals-fate-of-russian-academy-of-sciences-1.13785.

Schweitzer, Glenn E. *Containing Russia's Nuclear Firebirds: Harmony and Change at the International Science and Technology Center*. Athens: University of Georgia Press, 2013.

———. *DMZ: The Story of the International Effort to Convert Russian Weapons Science to Peaceful Purposes*. New York: M. E. Sharpe, 1996.

———. *Experiments in Cooperation: Assessing U.S.-Russian Programs in Science and Technology*. New York: Twentieth Century Fund Press, 1997.

———. *Interacademy Programs between the United States and Eastern Europe 1967–2009: The Changing Landscape*. Washington, DC: NAP Press, 2009.

———. *Scientists, Engineers, and Track-Two Diplomacy: A Half-Century of U.S.-Russian Interacademy Cooperation*. Washington, DC: National Academies Press, 2004.

———. *Swords into Market Shares: Technology, Economics, and Security in the New Russia*. Washington, DC: Joseph Henry Press, 2000.

———. *Techno-Diplomacy: US-Soviet Confrontations in Science and Technology*. New York: Plenum Press, 1989

Science and Engineering Indicators 1987. NSB 87-1. Washington, DC: National Science Board, 1987. Accessed December 26, 2018. https://babel.hathitrust.org/cgi/pt?id=osu.32435021892393;view=1up;seq=5

Science, Technology, and American Diplomacy 1995. Sixteenth Annual Report Submitted to the Congress by the President Pursuant to Section 503(b) of Title V of Public Law 95-426. Washington, DC: US Government Printing Office, 1996.

"Scientific Exchanges and U.S. National Security." *Science* 215, no.429 (January 8, 1982): 139–141.

Senechal, Marjorie. "Adventures of an Amateur Crystallogpher." *ACA History* (American Crystallographic Association), 2013. Accessed June 16, 2016. http://www.amercrystalassn.org/h-senechal_memoir.

———. "8 marta 1979: Women's Day in the Soviet Union." *OUPblog: Oxford University Press's Academic Insights for the Thinking World*. Accessed June 16, 2016. http://blog.oup.com/2014/03/international-womens-day-soviet-union/.

Sevostyanov, Nikolay. "Washington's Spies Are Getting Pretty Brazen." *Segodnia.Ru*, May 15, 2016. Accessed August 27, 2016. http://www.segodnia.ru/content/175750.

Sessler, Andrew M. "Physicists and the Eternal Struggle for Human Rights." *APS News* 4, no. 9 (October 1995). Accessed June 23, 2015. http://www.aps.org/publications/apsnews/199510/human-rights.cfm.

Sher, Gerson S. *Praxis: Marxist Criticism and Dissent in Socialist Yugoslavia*. Bloomington: Indiana University Press, 1977.

———. "Russia, Ukraine and Sanctions: A Double-Edged Sword." *International Business Times*, March 6, 2014. http://www.ibtimes.com/russia-ukraine-sanctions-double-edged-sword-1559910.

———. "Science Diplomacy and Beyond." Letter to *Science* 345, no. 6197 (August 8, 2014): 631.
———. "U.S.-Russia Scientific Cooperation in Changing Times." *Problems of Post-Communism*, July–August 2004: 25–33.
———. "Why Should We Care About Russian Science?" *Science* 289, no. 5478 (July 21, 2000): 389.
Sher, Gerson S. (with James D. Watson). "Does Research in the Former Soviet Union Have a Future?" *Science* 264, no. 5163 (May 27, 1994): 1280–81.
"Siberian financial whizz-kid appointed to lead Russian science." *Siberian Times*, October 25, 2013. Accessed June 5, 2018. http://siberiantimes.com/science/profile/news/siberian-financial-whizz-kid-appointed-to-lead-russian-science/.
Sinelschikova, Yekaterina. "Dismissal of American Vice Rector in Nizhny Novgorod Causes Controversy." *RBTH* (*Russia Beyond the Headlines*, an online publication of *Rossiyskaya gazeta*), July 3, 2015. Accessed August 27, 2016. https://rbth.com/society/2015/07/03/dismissal_of_american_vice_rector_in_nizhny_novgorod_causes_controver_47451.html.
Smith, Richard J. *Negotiating Environment and Science: An Insider's View of International Agreements, from Driftnets to the Space Station*. Washington, DC: Resources for the Future, 2009.
Soros, George. "Who Lost Russia?" *New York Review of Books*, April 13, 2000. Accessed June 20, 2015. www.nybooks.com/articles/archives/2000/apr/13/who-lost-russia?pagination=false&printpage=true.
Soviet Military Power. Series published from 1981 through 1990. Washington, DC: US Government Printing Office.
Soyfer, Valery. "Ne nado plevat' v ruku daiushchego" ["You must not spit into the donor's hand"]. *Troitskiy variant*, no. 184: 4–6. Accessed July 1, 2016. http://trv-science.ru/2015/07/28/ne-nado-plevat-v-ruku-dayushchego/.
———. *Vlast' i nauka: Razgrom kommunistami genetiki v SSSR* [Power and science: The destruction of genetics in the USSR by the communists] Moscow: CheRo, 2002.
Stalin, Joseph V. *History of the Communist Party of the Soviet Union (Bolsheviks): Short Course*. Moscow: Foreign Languages Publishing Press, 1939.
"Statement: The Russell-Einstein Manifesto." July 9, 1955. Pugwash Conferences on Science and World Affairs. Accessed June 18, 2015. http://pugwash.org/1955/07/09/statement-manifesto.
Stokes, Donald E. *Pasteur's Quadrant: Basic Science and Technological Innovation*. Washington, DC: Brookings Institution Press, 1997.
Stone, I. F. "The Sakharov Campaign." *New York Review of Books*, October 18, 1973. Accessed August 16, 2016. http://www.nybooks.com/articles/archives/1973/oct/18/the-sakharov-campaign/?pagination=false&printpage=true.
Stone, Richard. "A Painful Cure for Ailing Academy. *Science* 342, no. 6163 (December 6, 2013): 1157.
———. "Russia's Science Reform Czar on U.S. Sanctions List." *Science Insider*, March 21, 2014. Accessed April 21, 2017. http://www.sciencemag.org/news/2014/03/russias-science-reform-czar-us-sanctions-list.
———. "Embattled President Seeks New Path for Russian Academy." *Science*, February 11, 2014. Accessed December 20, 2018. https://www.sciencemag.org/news/2014/02/embattled-president-seeks-new-path-russian-academy
Strengthening U.S.-Russia Cooperation on Nuclear Nonproliferation: Recommendations for Action. Washington, DC: National Academies Press, 2005. Accessed May 19, 2015.

https://books.google.com/books?id=XMuaAgAAQBAJ&printsec=frontcover#v=onepage&q&f=false.

"Text of Lacy-Zaroubin Agreement, January 27, 1958." *Passing Through the Iron Curtain: Librarians and the U.S.-Soviet Cultural Exchange, 1950–1965*. Accessed July 12, 2015. https://librariesandcoldwarculturalexchange.wordpress.com/text-of-lacy-zaroubin-agreement-january-27-1958/.

"Thanks for Responding to 2010 Commercialization Survey." *US Industry Coalition E-Notes* 12, no. 2 (September 2011).

Tucker, Robert C., ed. *The Marx-Engels Reader*. New York: Norton, 1972.

Turchin, Valentin. "Boycotting the Soviet Union." *Bulletin of the Atomic Scientists* 34, no. 7 (September 1978): 7–11.

Twilley, Nicola. "How the First Gravitational Waves Were Found." *The New Yorker* (February 11, 2016.) Accessed December 15, 2018. https://www.newyorker.com/tech/annals-of-technology/gravitational-waves-exist-heres-how-scientists-finally-found-them.

The Unique U.S.-Russian Relationship in Biological Sciences and Biotechnology: Recent Experience and Future Directions. Report of the National Academy of Sciences. Washington, DC: National Academies Press, 2013.

USIC E-Notes, vol. 12, no. 2 (September 2011).

Wagner, Caroline S. *The New Invisible College: Science for Development*. Washington, DC: Brookings Institution Press, 2008.

Wagner, Caroline S., Irene T. Brahmakulam, D. J. Peterson, Linda Staheli, and Anny Wong. *U.S. Government Funding for Science and Technology Cooperation with Russia*. Santa Monica, CA: RAND Corporation, 2002. http://www.rand.org/pubs/monograph_reports/MR1504.html. Also available in print form.

Wolfe, Audra J. *Competing with the Soviets: Science, Technology, and the State in Cold War America*. Baltimore: Johns Hopkins University Press, 2013.

Wolin, Sheldon. *Politics and Vision*. Princeton, NJ: Princeton University Press, 1960.

INDEX

General topics are indexed under the heading "US-FSU scientific cooperation." Page numbers in italics refer to illustrations.

Academy of Sciences of the USSR (ASUSSR). *See* Russian Academy of Sciences; Soviet Academy of Sciences
Afghan-Soviet war, 31, 33–34, 80, 176–77
Agreement on Peaceful Uses of Atomic Energy, 34–35
Agreements in Science and Technology (US-Russia), 34–44
Agreements in Science and Technology (US-USSR): overview, 23–25, 24t; Afghan-Soviet war and, 33–34; Asian programs as precursor, 21–22; collaborative structure, 28–29; era of détente and, 20–21; House hearings on, 25–27; intergovernmental exchange as key feature, 16, 27–28, 31; program cancellation issues, 35
agriculture: DuPont collaboration, 134–35; FSU seed libraries, 110, 134–35; Nixon-Brezhnev agreements on, 23–25, 24t; Soviet withholding of data and, 26
Ailes, Catherine, 27, 29n2, 31–33
Akayev, Oskar, 63–64
Al'ferov, Zhores, 161
Alfimov, Mikhail, 64
Alfred P. Sloan Foundation, 48–49
Alkakhatsi, Maia, 119, 134, 157
American Astronomical Society (AAS), 48–49, 56
American Chemical Society, 56
American Mathematical Society (AMS), 48–49, 56
American Physical Society (APS), 48–49, 56
American Society for Microbiology, 56
anti-Semitism, 31, 76, 79, 176–77, 201–207, 246–47
Armenia, 66, 67t, 207

Artsimovich, Lev, 82
Assyria, 207
astronomy, 45
atmospheric sciences, 86–88
atomic energy: Agreement on Peaceful Uses of Atomic Energy, 34–35; collaboration in, 93–94; Nixon-Brezhnev agreements on, 23–25, 24t; US-Russia Agreements, 43–44; USSR uranium concerns, 97–98; US tritium and plutonium processing, 109. *See also* physics; weapons of mass destruction
Atoms for Peace Program, 11
Azbel, Mark Ya., 246–47
Azerbaijan, 66, 67t

Baker, James, 50
Balzer, Harley, 64, 160
Barghoorn, Frederick, 104, 112n56, 215n9
Basic Research and Higher Education in Russia (BRHE), 64–66, 84, 132–33, 159–62, 213–14, 270
Bayev, Aleksandr, 204–205
Belarus, 50, 67t
Bell, David, 107–109, 134–36, 183–85
Bell Labs, 190
biology: anthrax and bioweapon concerns, 98–99, 118, 128–29, 165–66, 211–12; balance of program benefits, 263; botany, 86–88, 119, 157; Eliava Institute, 117–19, 128–29; Institute of Molecular Genetics, 215n24; Soviet international stature in, 45; USSR bio-industry, 197
Biotechnology Engagement Program (BTEP), 147, 197–98
Boeing Corporation, 96, 108, 115–16, 124–25, 155–56, 163n32, 180–81

295

Bolotin, Sergei, 235
botany, 86–88, 119, 157
Bozorth, Richard, 246–47
Braginsky, Vladimir, 140–43, 206–207, 212, 223
Brands, H. W., 37
Brazil, 173, 178
Brezhnev, Leonid M., 16, 20–22
Brigham-Grette, Julie, 89–90, 110, 114, 144–46, 199, 242–45
Bromley, D. Allan, 46–47, 269–70
Bronk, Detlev, 16
Brookhaven National Laboratory, 59–60
Brown, George E., Jr., 50, 53, 62
Brown, Harrison, 22
Budker Institute of Nuclear Physics, 204
Bulletin of the Atomic Scientists, 10, 18
Burt, Richard, 36
Bush, George H. W., 48
Bush, George W., 163n32
Byrnes, Robert, 17

Cairo Initiative, 66
Campbell, Cathleen, 103
Canada, 58
Carey, William, 116–17
Carlucci, Frank, 116–17
Carnegie Corporation, 65
Carter, Ashton, 99
Carter, Jimmy, 23, 31, 33–34, 39n2, 177
Centers for Disease Control (CDC), 88–89
Central Intelligence Agency (CIA), 210–211, 223
CERN (the European Organization for Nuclear Research), 44–46
chemistry: chemistry collaboration, 90–92; DuPont collaboration in, 184; petroleum delegation of 1961, 101; Soviet international stature in, 45
China, 11, 22–23, 173, 178
citizen diplomacy, 12–13, 20
Civilian Research and Development Foundation (CRDF): overview, 62–70, 67t; AIDS and Hepatitis C research, 146–47; Bell award from, 108; BRHE program, 64–66, 84, 132–33, 159–62, 213–14, 270; Cathleen Campbell role, 103; closing of Moscow office, 68; commercial accomplishments and, 182–83; competitive grant mechanism, 161; corruption issues, 213–14; defense industry research and, 252–54; equipment and instrumentation initiatives, 120; mutuality of benefit principle, 269; Next Steps to the Market Program, 64, 108, 135, 184; program oversight and, 191; Randolph Guschl role, 109; Regional Experimental Science Centers, 156
climate science, 144–46, 242–45
Clinton, Bill, 63, 155
Cohen, Stephen F., 43, 103, 105, 133, 270, 274
cold-weather construction, 154, 180
Collins, James, 124
Colwell, Rita, 114, 192
commercial motives and accomplishments: commercial accomplishments, 179–85; congressional support and, 181–82; GIPP collaborative design and, 182–84; nonproliferation programs and, 135–36, 181–82; US entrepreneurial interests, 105–106, 133–34; US Industry Coalition, 60, 108, 135–36, 172–73, 182–83
Committee on Exchanges (COMEX), 210–211
Communist Party of the Soviet Union (CPSU), 11
computers. *See* information technology
construction: cold-weather construction, 154, 180; Nixon-Brezhnev agreements on, 24t
CRDF Global (US Civilian Research and Development Foundation), 50–51, 57, 66–68
Crimea, 9, 31, 69, 175–76, 203–204
Crum, Lawrence, 84–85, 114, 130, 158
Cuba, 123–24, 137

Danilenko, Vyacheslav, 166–67
David, Edward, Jr., 21–22, 29n3, 177
David-Fox, Michael, 19n12
Davydov, Alexander Sergeevich, 91–92

Delone, Boris Nikolayevich, 148
Deng Hsiao Ping, 23
détente: arms control and, 124; intergovernmental structures and, 28, 31, 188–89; Nixon-Brezhnev meeting and, 20–21; Nixon-Kissinger political doctrine and, 52; Soviet "applied microbiology" and, 30n17
Dezhina, Irina, 45–46, 48, 54, 64, 159, 161, 170, 220, 268
d'Herelle, Felix, 118
Drever, Ron, 141
Duffy, Gloria, 260
DuPont Corporation, 109–110, 134–35, 184
Dykhne, Aleksandr, v, xi, 66
Dynasty Foundation, 239n31
Dzerzhinsky, Feliks, 103–104

earth sciences. *See* geosciences
Eaton, Cyrus, 10
Ehrlich, Paul, 157
Einstein, Albert, 10
Eisenhower, Dwight D.: "Atoms for Peace" exhibit, 81; citizen diplomacy program, 11–13, 20, 23, 28, 52, 272–73; mutual understanding initiative, 259; promotion of democracy and, 131; Russian national exhibition exchange, 101
Eisenhower, Susan, 39
Eliava Institute of Bacteriophages, Microbiology, and Virology, 117–19, 128–29
energy development, 43–44. *See also* atomic energy
Environmental Protection Agency, 24t, 62, 103
environmental science: climate science, 144–46, 242–45; ecological diversity of Georgia and, 120; exemption from 1982 termination, 34; Nixon-Brezhnev agreements on, 23–25, 24t; Reagan environmental policies and, 34–35; US-Russia Agreements, 43
Estonia, 207, 232
European Union, 50, 58

Federal Agency for Scientific Organizations (FASO), 231
Federal Bureau of Investigation (FBI), 210–212, 223, 265
Federal Security Service, 253
Fedorov, E. P., 86
Fermilab, 204
Feynmann, Joan, 262
Finarelli, Margaret, 155
Finland, 58
foreign policy (US). *See* US foreign policy
Fowler, Max, 94
France, 10, 44–45, 51, 204–205
Frank-Kamenetskii, Maxim, 204–205, 214
FSU (former Soviet Union), 5n4. *See also* Russian Federation; Soviet Union
Fulbright, J. William, 130
Fulbright scholarship program, 130, 210
Fursenko, Andrey A., 65, 72n48, 160, 164n43, 231, 239n21

Galiulin, Ravil, 148
Gardner, David, 190
Gel'fand, Israel, 205
Georgia: AIDS and Hepatitis C research, 146–47; botany research in, 88, 119; CRDF in, 66, 67t, 146–47; ecological diversity, 120; Eliava Institute, 117–19, 128–29; human rights issues, 207; Ilia State University, 233–34; science reform, 233–34
geosciences: balance of program benefits, 263; Bering Land Bridge, 89–90; evolution theory, 264; Hearn training in, 102; intelligence gathering and secrecy concerns, 211–12; Lake El'gygytgyn coring expedition, 144–46, 242–45; "location" as collaboration motive, 114–15; Peter Raven contributions, 86–88; Soviet geology, 137–38n3
Germany, 219, 221, 264
Gilot, Francoise, 250–51
Ginzburg, Vitaly, 76–77, 202, 223
Givargizov, Yevgeniy, 147
glasnost (open discussion), 78

Global Initiatives for Proliferation Prevention (GIPP): overview, 60–61; commercial accomplishments and, 182–84; congressional skepticism and, 175; DuPont participation in, 134–36; Gottemoeller role in, 100; Lowe role in, 152–53; scientific professionalism and, 172–73
Goldfarb, Alexander, 53, 157
Goldsmith, Hyman, 10
Gonchar, Andrey, 55, 161
Gorbachev, Mikhail, 37–38, 39, 78, 203
Gore, Al, 63
Gore-Chernomyrdin Commission, 44
Gottemoeller, Rose, 100, 127, 155
Graham, Loren: assessment of program accomplishments, 260; BRHE program and, 159–61, 214; CRDF funding and, 64; on ISF support, 268; as Kaysen panelist, 266; peace ideal as personal motivation, 130, 132–33; Russian language study, 92–93; on science and democracy, 270; as Sher influence, xi–xii; on Soviet corruption/inefficiency, 214, 220; on Soviet geology, 137–38n3; on US-Soviet cultural differences, 264
Graham, William R., 38
Grant Assistance Program (GAP), 56–57, 66
Guschl, Randolph, 109–110, 134–36, 183–84

Halem, Claude, 204–205
Handler, Philip, 33
Handley, F. Gray, 198
Hawking, Stephen, 240–42
health sciences: AIDS/HIV and Hepatitis C research, 146–47, 249–50; exemption from 1982 termination, 34; idiosyncrasies of FSA science and, 199–201; Nixon-Brezhnev agreements on, 23–25, 24t; smallpox eradication collaboration, 88–89; US-Russia Agreements, 43
Hearn, Paul, 102–103, 113–15
Hecker, Siegfried: governmental program oversight and, 190–91; high regard for Soviet scientists, 93–94; nonproliferation collaboration and, 68–69, 94–95, 149–50, 165–67, 171; on US and FSU scientific styles, 218; WMD concerns and, 48, 128

Hefter, Heinz, 109
Hoffman, David, 97, 128
Hoffman, Roald, 90–92, 136–37, 259
Holgate, Laura, 99, 127–28, 166, 170–72
Hopkins, David, 89
human rights: anti-Semitism, 31, 76, 79, 176–77, 201–207, 246–47; as Carter administration emphasis, 31, 39n2; collaboration issues with, 199–208; International Women's Day and, 248–49; Sakharov persecution, 31, 33, 77, 93–94, 113, 176, 201–203, 274; Shcharansky trial, 31–32

India, 173, 178
Industrial Partners Program (IPP), 59–61
information technology: comparison of scientific styles and, 218–19; computer acquisition, 48, 119, 156–57; geology applications, 137–38n3; network development, 56; parallel processing development, 149, 181
Initiatives for Proliferation Prevention (IPP), 60–61, 110
Institute of Chemical Physics, 151
Institute of Crystallography, 248–49
Institute of International Education, 12
Intel, 181
intelligence gathering and secrecy: overview, 208–213; CIA and, 210–211, 223; closed cities, 47, 60, 95, 98, 168, 190; debriefing requirements, 209, 211, 263; DOD concerns about, 116–17, 138n7, 263; FBI and, 210–212, 223, 265; KGB intervention in scientific meetings, 245–46; open and closed societies, 1, 265, 267; Soviet political control of science, 225–27; Soviet scientist nonproliferation deniability, 252–54. See also KGB
interacademy exchange programs: overview, 16–18; Afghan-Soviet war sanctions, 33–34; initial agreement for, 16; institutional problems with, 188; intergovernmental agreements and, 29n2; lab-to-lab programs, 48–49, 59, 190; Soviet Union dissolution and, 44
intergovernmental agreement of 1987, 38–39

intergovernmental structures, 28, 31, 188–89
International Association for the Promotion of Cooperation with the Scientists of the Former Soviet Union (INTAS), 51, 161, 268
International Atomic Energy Agency (IAEA), 99, 166
International Center for Electron Beam Technology (ICEBT), 108, 157
International Conference on Nonlinear Acoustics, 84
International Monetary Fund (IMF), 53
International Research and Exchanges Board (IREX), 15–17, 90–92, 105, 188, 210
International Science and Technology Center (ISTC): overview, 57–59, 168–69; assistance vs. cooperative model and, 268, 273; commercial accomplishments and, 181–82; founding, 50; funding structure, 197; program decline and closure, 173–75
International Science Foundation (ISF): overview, 52–57, 56t; assistance vs. cooperative model and, 267–69, 273; Braginsky research team support, 143–44; "brain drain" concerns and, 124; competitive grant mechanism, 161; corruption issues, 214; equipment and instrumentation initiatives, 120, 157; founding, 51; funding structure, 197; peer-review initiative, 132
International Women's Day, 248–49
Inter-University Committee on Travel Grants (IUCTG), 15–17
Iran, 166–67
Isaacson, Richard, 141–43

Japan, 21–22, 50, 58, 178
Javakishvili, Zurab, 234
Jewish Soviet researchers, 32–33
John D. and Catherine T. MacArthur Foundation, 64–65
John Paul II (Wojtiła, Karol), 34

Kadanoff, Leo, 80
Kaganov, Moisiei, 208–209, 245–47
Kahn, Louise Wolf, 83
Kapitsa, Peter, 222

Kasha, Michael, 90–91
Kassof, Allen, 104
Kaysen, Carl, 18
Kazakhstan, 59, 67t, 97–98, 129–30, 165–66
Kennedy, John F., 21, 36, 112n56, 125
KGB (Committee for State Security), 80, 104, 194, 205, 212, 226, 245–46, 253. See also intelligence gathering and secrecy
Khrushchev, Nikita, 11, 15, 101
Kiser, John, 106–107, 180
Kissinger, Henry: China exchange, 22–23; as collaboration advocate, 20, 21–22, 29n2; foreign policy principles, 23, 122; "incentives for restraint" approach, 20, 23, 25, 34, 39, 122, 176–79, 259, 274; SALT treaty and, 21
Knapp, Edward, 93
Komarov Botanical Institute, 87–88, 157
Kotyukov, Mikhail, 232
Kozlov, Valery, 232
Krosik, Yakov, 184–85
Kuban' State University, 213–14
Kurchatov, Igor, 81
Kutter, Elizabeth, 119
Kyrgyzstan, 63–64, 67t

Lacy, William S. B., 13
Lacy-Zaroubin Agreement, 13–17, 110n4, 188, 261
Landau, Lev Davidovich, 221–23
Landau School (Soviet science), 221–23
Lane, Neal, 63, 252
language and culture: communication issues and, 121, 192–93, 223; Holiday Inn incident, 254–55; language and culture junkies, 97, 100–105; new words related to collaboration, 49; Russian scientific language, 227–29; science as a common language, 1–2, 217; *Ugly American* stereotype and, 12–13; US scientists embrace of, 92–93
Laser Interferometer Gravitational-Wave Observatory (LIGO), 141, 143, 201
Latvia, 207, 232
Lawrence Livermore National Laboratory, 47, 59–60, 114
Leontovich, Mikhail, 82

Lifshitz, Ilya Mikhailovich, 208–209, 245–47
Lifshitz, Yevgeny, 77
Lincoln, Abraham, 17
Lithuania, 207, 232
Livanov, Dmitri, 160
Lockheed Martin, 163n32, 190
Logunov, Anatoliy, 206–207
Lowe, Terry, 94–95, 117, 136, 152–54, 183
Lugar, Richard, 175–76, 271
Lviv Polytechnical Institute, 208
Lysekoism, 30n17

Malloy, Eileen, 95–96, 173–74, 191–94, 250–52
Manhattan Project, 10
manufacturing: Soviet *bulat* technology, 108; US patents for Soviet technology, 106–107
Markusova, Valentina, 161, 162
Mashko, Sergey, 252–54
Maslow, Marvin, 183
materials and metallurgical sciences: cold-weather construction, 154, 180; Institute of Crystallography, 248–49; nanotechnology, 94–95, 152–54, 183; plasma acceleration research, 184–85; Terry Lowe nonproliferation work in, 94–95; Ukraine welding research, 108, 179–80, 207; US patents for Soviet technology, 106; US-USSR Working Group on Electrometallurgy and Materials, 158
Materials Research Society (MRS), 94–95
mathematics: balance of program benefits, 263; crystal modeling collaboration, 83–84; freedom of travel and, 76–77; mathematical "Olympiads," 224–25; nonlinear acoustics collaboration, 84–85; Soviet international stature in, 45, 77; Soviet mathematician career changes and, 41; topology of crystals research, 147–49
Mathisen, Ole, 92
Matlock, Jack, 36
McFarlane, Robert ("Bud"), 36
medical science and health. *See* health sciences
Medvedev, Dmitriy, 160
Meshkov, E. E., 149
Metallicum, 152, 183
microbiology, 30n17, 211–12

Microsoft, 181
Ministry of Atomic Energy, 47
Mirov, Nicholas, 86
Mishchenko, M., 235
Misner, Charles, 241
Missouri Botanical Garden, 88
Mitchell, Lawrence, 18, 105
Mitterand, François, 44–45, 51
Moldova, 66, 67t
Moscow State University, 77, 140, 142–44, 206
Movchan, Boris, 157–58, 236–37
Multi-Arc Vacuum System Inc., 108

nanotechnology, 94–95, 152–54, 183. *See also* materials and metallurgical sciences; physics
Narath, Al, 190
National Academy of Sciences (NAS): Bromley study, 46–47, 269–70; China engagement and, 22; intelligence gathering policies, 209–210; interacademy program development, 16–18; ISF grants and, 56; mathematics collaboration and, 84; Okun' advocacy, 32–33; petroleum delegation of 1961, 101; refusenik crisis and, 202–203; Sher employment, 105; Shulman role in, 111n34; Soviet Union dissolution and, 51, 168–69
National Academy of Sciences of Ukraine, 85–86, 234–37
National Aeronautics and Space Administration (NASA), 155, 212, 227
National Institute for Allergy and Infectious Diseases (NIAID), 189
National Institutes of Health (NIH), 88, 147, 189, 196–97, 269
National Nuclear Security Administration (NNSA), 136, 173
National Science Foundation (NSF): awards involving Russia, 69t; Cooperative Grant Program, 252–53; CRDF oversight and, 63; earth sciences collaboration, 89; funding structure, 195–97; intelligence gathering and secrecy concerns, 117, 138n7, 211; interacademy program funding, 16; ISF compared to, 56; Japan cooperative

program and, 21; lab-to-lab programs support, 48–49; LIGO funding requests, 141, 143; as Nixon-Brezhnev agreement executive agency, 25, 27–28; oversight of collaborative programs, 189–90; US participant grants, 269; US-Russia Agreements, 44; US-USSR Agreements, 25
National Security Council (NSC), 14, 38, 100, 127, 150, 176
Nazarbayev, Nursultan, 98
Nesmeyanov, A. N., 16
Neureiter, Norman, 21–23, 97, 100–101, 125–26, 177
Nigeria, 88–89
Nixon, Richard, 15–16, 20–22, 101
Nixon-Brezhnev intergovernmental exchange programs. See Agreements in Science and Technology (US-USSR)
Nizhniy Novgorod State University (NNGU), 271
nonproliferation programs: overview, 57, 165–69; assessment of impact and success, 268–69, 271–73; bureaucratic and logistical issues, 173–75, 199, 215n14; commercial motives and accomplishments, 135–36, 181–82; funding structure, 197; Hecker collaborations, 94–95; IPP/GIPP programs, 59–61, 100, 110; ISTC, 57–59, 168–69; Nunn-Lugar nonproliferation program, 99, 127–28, 167, 175–76, 271; people-oriented nonproliferation, 168–72; Project Sapphire, 98, 165–66; scientific professionalism and, 172–73; STCU programs, 57–59; Weber contributions, 97–98. See also Soviet Union dissolution; weapons of mass destruction
North Korea (Democratic People's Republic of Korea), 128
Norway, 58
Novikov, Igor, 76–77
Nuckolls, John, 48, 150, 190
nuclear energy. See atomic energy; physics
Nunn, Sam, 175–76, 271

Oak Ridge National Laboratory, 59–60, 98
Obama, Barack, 66, 99
ocean studies, 43

Okun,' Lev B., 32–33, 44–45, 49–53, 205–206
Olympic Games boycott of 1980, 33–34, 177
Oparin, Alexander, 264
Open Society Institute, 54, 66
Oppenheimer, Robert, 10
O'Rourke, P. J., 163n32
Osipov, Yuriy, 50
Owens, Charles T. "Tom," 63
Ozernoy, Leonid, 202

Palmer, Mark, 36
Pardee, Arthur E., Jr., 27, 29n2, 31–33
Paton, Borys Yevhenovych, 234–37
Paton Electric Welding Institute, 108, 179–80, 207, 235
Penrose, Roger, 240
People-to-People Program, 11–13, 20, 23, 28, 272
perestroika (rebuilding), 78, 103, 177, 203, 229
Perry, William, 99
Peter the Great, 17
Petrov, Rem, 78
Phygen Coatings Inc., 107–108, 134–35, 184–85
physics: balance of program benefits, 263; Fermilab collaboration, 204; freedom of travel and, 76–77, 142; gravitational wave detection research, 140–44; Hawking Radiation discovery, 240–42; hydrodynamics, 211–12; intelligence gathering and secrecy concerns, 117, 138n7, 211–13; Landau School physics, 222–23; Soviet international stature in, 45, 77; working group in physics meeting cancellation, 32–33. See also atomic energy; nanotechnology
Pickering, Thomas, 95–96, 115–16, 122–24, 178, 180
Pines, David, 32–33
Pioneer Seed, 110
Poland, 31, 34–35, 176–77, 233
Popper, Karl, 132
Poroshenko, Petro, 236
Proctor, Michael, 121
Project Sapphire, 98, 165–66
public health. See medical science and health

Pugwash Conferences on Science and World Affairs, 10–11, 101
Putin, Vladimir: defense industry research and, 273; Dykhne telegram, v; higher education consolidation initiative, 159–61; hostility to cooperation and, 272; nonproliferation assistance and, 173; rejection of foreign NGOs, 239n31; restrictions on international publication, 225–26; university-based research and, 273–74; on the USSR dissolution, 41; view of US assistance, 59

Rabinowitch, Eugene, 10–11
Rabinowitch, Victor, 64–65
Raven, Peter, 63, 86–88, 114, 120, 130, 134, 157, 227–29
Reagan, Nancy, 36
Reagan, Ronald: Afghan-Soviet war sanctions, 34; collaboration philosophy of, 35–37; environmental policies, 34–35; "evil empire" speech, 16, 35, 37; foreign policy experience, 130; "trust but verify" approach, 191
refuseniks, 31, 32–33, 201–202
Reischauer, Edwin O., 21
Richmond, Yale, 13–14, 267
Roberts, Paul, 121
Rogers, William, 212
Rubbia, Carlo, 44–46, 49–53
Russell, Bertrand, 10
Russell-Einstein Manifesto, 10
Russian Academy of Sciences (RAS): CRDF development and, 64–65; founding, 17; RAS building repair, 157; Rubbia-Okun' funding initiative and, 50; science reform and, 230–35; scientific output study of, 162. See also Soviet Academy of Sciences
Russian Federation: Rubbia letter references to, 45; scope of "Russia" term, 5n4; university-based research, 233–34, 270–71, 273–74. See also FSU; Soviet Union; Soviet Union dissolution
Russian Foundation for Basic Research, 44, 161, 231
Russian language. See language and culture
Russian Ministry of Education and Sciences, 65
Russian Studies, 104

Ruzmaikin, Alexander, 121, 226–27
Ryabchenko, Sergiy, 237

Sagdeev, Roald Z., 32, 39n3, 80–82, 220, 222, 229
Sakharov, Andrey, 31, 33, 77, 93–94, 113, 176, 201–203, 274
Salk, Jonas, 250–51
Saltykov, Boris, 64, 230–31
Sauer, Andrew, 121
Schweitzer, Glenn, 58–59, 64, 96, 126, 168–69, 173, 181–82, 202–203
science: advancement of knowledge goal, 273; democracy effect on, 270; moral responsibility in, 11; science as a common language, 1–2, 217; scientific professionalism as collaboration motivation, 113–15, 171–73
Science and Technology Center Ukraine (STCU), 57–59, 182, 197–98, 273
Senechal, Marjorie, 82–84, 114, 147–49, 160, 224–25, 248–49
Serbin, Ivan, 82
Sevostyanov, Nikolay, 271
Shcharansky, Anatoly (Natan Sharansky), 31–32
Sheftal, Nikolai Naumovich, 83, 147
Shemyakin Institute of Bio-Organic Chemistry, 211–12
Sher, Gerson S., 18–19, 38, 63–64, 104–105, 131, 252–55
Shklovskii, Boris, 79–81, 203–204, 222, 224
Shklovskii, Iosif, 76, 79
Shulman, Marshall, 91, 111n34
Shultz, George, 38
Simons, Thomas, 36
Skvortsov, Alexei, 87–88
Smith, Richard J., 38
Solomon, Anne (Keatley), 22
Solomonia, Revaz, 225, 233–34
Soros, George: assistance vs. cooperative model and, 267–69; Braginsky research team support, 143–44; CRDF support, 63; funding structure, 197; GAP program and, 66; ISF overview, 52–57; ISF Soviet dissolution strategy, 124; Komarov Botanical Institute support, 88; lab-to-lab

programs and, 48; open society initiative, 132; Rubbia-Okun' funding initiative and, 51; science advocacy, 259

South Korea (Republic of Korea), 31, 35, 58, 176

Soviet Academy of Sciences (ASUSSR): fake award incident, 251–52; Ioffe Physico-Technical Institute, 79, 161; Komarov Botanical Institute, 87–88, 157; NAS-ASUSSR interacademy program, 16–18; role in Stalin regime, 219–20, 222; Senechal work at, 147; Theoretical Physics working group meeting and, 32–33; US ASUSSR journal readers, 83; US-USSR S&T agreements and, 24t. *See also* Russian Academy of Sciences

Soviet Nuclear Fission (Holgate and Carter), 99

Soviet State Committee for Science and Technology (SCST), 25

Soviet Union: closed society model, 1, 265, 267; international exchange programs, 19n12; isolation from world science, 223; Marxist philosophy and, 104–105, 264; "mirror image" foreign policy hypothesis, 215n9; motives for collaboration, 117–22; persecution of scientists, 118; quality of scientific research, 115, 137; research freedom, 115–16, 137; research organizational structure, 151; science reform, 229–38; scientists and engineers data, 70n3; scope of term, 5n4; Soviet science overview, 219–20; Soviet science "schools," 221–23; top-down science organizational structure, 42, 45, 49, 115, 219–21, 229, 271; US patents for Soviet technology, 106–107. *See also* FSU; Russian Federation

Soviet Union dissolution: ASUSSR and, 17; "brain drain" concerns, 41–42, 46–48, 124, 159, 170, 235; emergency assistance programs, 46–47, 71n18, 267–72, 273; equipment and instrumentation initiatives, 120; impact on Braginsky research team, 140–44; "internal emigration" of scientists, 54–55; Lacy-Zaroubin Agreement and, 15; oversight of collaborative programs and, 189–90;

private cash payments to FSU scientists, 47, 49, 114; Rubbia letter and, 44–46, 49–53; science reform and, 41–42, 229–38; scientific infrastructure, 156–62; Soviet intergovernmental agreements and, 43–44; US democratization strategy, 14, 42–43, 46–47, 131–33, 269–70, 274; WMD concerns and, 97–98, 128, 149. *See also* nonproliferation programs; Russian Federation

Soyfer, Valery, 55, 157

space programs: Apollo-Soyuz Test Project, 26, 34; intelligence gathering and secrecy concerns, 212–13; intergovernmental agreement of 1987, 38–39; Nixon-Brezhnev agreements on, 23–25, 24t; political scrutiny of, 34; prejudice against Russian scientists, 262; research collaboration, 85–86, 163n32; space science overview, 154–56; *Sputnik*, 12–13, 105, 154–55, 265; US-Russia Agreements, 43; Yuri Gagarin space flight, 105, 112n59

Spirit of Pugwash, 10–11. *See also* Pugwash Conferences on Science and World Affairs

Sputnik, 12–13, 105, 154–55, 265

Stalin, Joseph, 11, 17, 219–20, 222, 225, 229, 232

Starobinsky, Alexei, 240–42

Stever, Guyford H., 26

Strategic Arms Limitation Treaty (SALT), 21

Strikhanov, Mikhail, 65

Stuntz, Mayo, 209–210

Sweden, 58

Tajikistan, 67t

Tediashvili, Marina, 117–19, 128–29

Terenin, A. N., 91

Thorne, Kip: gravitational wave detection research, 140–44; on international publication, 226; interview summary, 75–77; Lacy-Zaroubin Agreement and, 188; rapport with Soviet scientists, 113, 206–207; refusal of debriefing by, 263; Sakharov advocacy, 201–202; on Soviet science "schools," 222–23; Stephen Hawking Moscow trip and, 240–42

Tikhonov, Aleksandr, 65

Topchiev, Aleksandr V., 101, 125–26
Tsertsvadze, Tengiz, 78–79, 146–47, 194
Tucker, Robert C., 103
Turkmenistan, 67t

Ufa State Aviation Technical University, 94, 152
Ugly American, The, 12–13
Ukraine: CRDF in, 50, 66, 67t; human rights issues, 207; International Center for Electron Beam Technology (ICEBT), 108, 157; Khar'kiv Institute of Physics and Technology, 221–22; National Academy of Sciences of Ukraine (NASU), 85–86, 234–37; post-Soviet "brain drain" in, 235–36; Russian annexation of Crimea, 9, 31, 69, 175–76, 203–204; science reform, 234–38; seed libraries, 110, 134–35; STCU in, 57–59, 182, 197–98, 273; technological advancement in, 107–108, 179–80
United Kingdom, 121
University of Minnesota, 80–81
US foreign policy: overview, 122–27, 176–79; collaboration philosophy and, 52; détente science policy, 20–21, 124; global security, 127–30; human rights policy, 31–32, 39n2; Khrushchev "peaceful coexistence" principle and, 11; Kissinger "incentives for restraint" approach, 20, 23, 25, 34, 39, 122, 176–79; Nixon exchange programs as tool, 23; nonproliferation programs and, 57–61; open society initiative, 132; Pickering "Hippocratic diplomacy" approach, 178; promotion of democracy, 14, 42–43, 46–47, 131–33, 269–70, 274; Reagan-era science-politics policy relationship, 39; "send messages" strategy, 28–29, 176–77, 274–75; world peace initiative, 130–31
US-FSU scientific cooperation—general headings: overall assessment, 274–77; bureaucratic and logistical issues, 173–75, 199, 215n14; collaborative publication, 169–70, 225; collaborative trust and, 150; complementarity, 217–18; corruption, 213–14; defections and emigrations, 113–14; diplomatic collaboration, 95, 126–27; equality and reciprocity principle, 261–62; freedom of travel and, 76–77, 79, 142, 250–51; funding structure, 194–98; interview strategy, 75; lack of understanding, 192–94; nonscientific goals and accomplishments, 165; peer-review system initiative, 132, 161; problems overview, 187; program impact and accomplishments, 259–61, 272–73; program inflexibility, 187–92; research freedom, 115–16, 137; research universities as hosts, 159–62; reverse technology transfer, 180; scientific accomplishments, 140; scientific infrastructure, 156–62; scientific professionalism as motivation, 113–15, 171–73; study methodology, 3, 4–5n3; timeline, 9; US patents for Soviet technology, 106–107

US-FSU scientific cooperation—chronology
—1940s: *Bulletin of the Atomic Scientists* (1945), 10
—1950s: Eisenhower "Atoms for Peace" initiative (1953), 81; Russell-Einstein Manifesto (1955), 10; Khrushchev "peaceful coexistence" speech (1956), 11; Eisenhower People-to-People Program (1956), 11–13, 20, 23, 28, 272; Pugwash Conference (1957), 10–11, 101; *Sputnik* (1957), 12–13, 105, 154–55, 265; Lacy-Zaroubin Agreement (1958), 13–17, 110n4, 188, 261; Geneva "Atoms for Peace" conference (1958), 81; Nixon-Khrushchev "kitchen debate" (1959), 15, 101; Soviet Academy of Sciences agreement (1959), 16; IUCTG/IREX program (1959), 15–17, 90–92, 105; US-Soviet national exhibition exchange (1959), 101; Yuri Gagarin space flight (1961), 105, 112n59
—1960s: Arms Control and Disarmament Agency (1961), 96; Soviet petroleum delegation (1961), 101; Topchiev lectures (1961), 125–26; Conferences on General Relativity and Gravitation (1965–1968), 76–78; Mega-Gauss conferences (1965), 94
—1970s: Sakharov human rights campaign (early 1970s), 9, 31–34, 176–77, 201–203,

274; Nixon-Brezhnev summit (1972), 16, 20–22; Strategic Arms Limitation Treaty (1972), 21; Agreements in Science and Technology, US-USSR (1972–1974) — *see main heading*; Apollo-Soyez docking (1975), 154–55; International Botanical Congress (1975), 87; House hearings on Nixon-Brezhnev agreements (1976), 25–27; Kiser report on Soviet US patents (1976), 106–107; Kaysen Panel Report (1977), 18, 263, 266; Shcharansky trial (1978), 31–32; China exchange agreement (1979), 22–23; Afghan-Soviet war (1979–1989), 31, 33–34, 176–77; Lake Sevan conference (1979), 80; working group in physics meeting cancellation (1979), 32–33
—1980s: Moscow Olympics boycott (1980), 33–34, 177; Poland martial law (1981), 31, 34–35, 176–77; termination of intergovernmental programs (1982), 28–29, 34–35; Reagan "evil empire" speech (1983), 35, 37; KAL 007 incident (1983), 31, 35, 176–77; Reagan address on S&T collaboration (1984), 36–37; Vega space research program (1984–1985), 85–86; Gorbachev *perestroika* period (1985), 78, 103, 177, 203, 229; National Security Study Directive 189 (1985), 138n7; International Conference on Nonlinear Acoustics (mid-1980s), 84; Reykjavik Summit (1985), 37–38; *Challenger disaster* (1986), 155, 163n32; *Mir* space station (1986), 155; renegotiated intergovernmental agreement (1987), 38–39
—1990s: Rubbia letter (1991), 44–46, 49–53; fall of the Berlin Wall (1991), 203; Livermore assistance initiative (1991), 47; Nunn-Lugar Cooperative Threat Reduction Act (1991), 99, 127–28, 167, 175–76; Soviet Union dissolution (1991) — *see main heading*; Saltykov science reform initiative (1991–1997), 231; ISTC proposal (1992), 50; Agreements in Science and Technology, US-Russia (1992–1994), 34–44; Bromley study (1992), 46–47, 269–70; European Materials Research Society (1992), 94; Agreements on Science and Technology Cooperation (1993), 25–26, 33; Foreign Operations, Export Financing, and Related Programs Appropriation Act (1994), 59; FREEDOM Support Act and CRDF (1995), 62–70, 67t; collapse of Russian ruble (1998), 65
—2000s: International Space Station (2000), 155
—2010s: Russia withdrawal from ISTC (2011), 175, 271; Russia annexation of Crimea (2014), 9, 31, 69, 175–76, 203–204; GIPP appropriation expiration (2014), 175; Nunn-Lugar program termination (2014), 271; Nunn-Lugar *New York Times* op-ed (2015), 175–76; Fortov forced resignation (2017), 231–32; RAS reorganization (2018), 232

US government entities: Arms Control and Disarmament Agency, 96; Central Intelligence Agency (CIA), 210–211, 223; Defense Intelligence Agency (DIA), 211; Department of Agriculture, 26; Department of Commerce, 265–66; Department of Defense (DOD), 63, 116–17, 129, 138n7, 191–92, 263; Department of Energy (DOE), 47, 59–60, 109–110, 127; Department of Health and Human Services (HHS), 66, 197; Department of Transportation, 26; Environmental Protection Agency, 24t, 62, 103; Federal Bureau of Investigation (FBI), 210–212, 223, 265; House Science and Technology Committee, 50, 62; Office of Technology Assessment, 62; Patent Office, 106–107; State Department (*see* US State Department); USAID (United States Agency for International Development), 157, 191–92; US Geological Survey, 102, 196–97; White House Science Office, 21, 29n3, 38, 62, 97, 103
US Industry Coalition (USIC), 60, 108, 110, 135–36, 172–73, 182–83
US-Russia Foundation, The, 271
US-Russian Expert Committee, 66

US State Department: as collaboration ally, 265–66; collaboration visa accommodation, 261; Cooperative Threat Reduction Program, 63–64; CRDF relationship, 63, 66; Foreign Service science programs, 96–97; intergovernmental agreement of 1987 and, 38; intervention in collaboration programs, 176, 189–90; Nixon-Brezhnev agreements and, 27; US-USSR S&T agrements and, 26
US-USSR Working Group on Electrometallurgy and Materials, 158
US-USSR Working Group on Special Topics in Physics, 32, 143
Uzbekistan, 67t

Valiev, Ruslan, 94, 152–54, 183
Vasilyeva, Ol'ga, 160
Vilenkin, Alexander, 222
VNIIEF (All-Union Research Institute of Experimental Physics), 93–94
VNIITF (All-Russian Scientific Research Institute of Technical Physics), 150
Voevodskaya, Marianna, 252–54

Wałęsa, Lech, 34
Watkins, James D., 48, 150
Waxmonsky Gary, 103–104
weapons of mass destruction (WMD): collaboration security benefits, 127–28; decommissioning procedures, 151; Khrushchev "peaceful coexistence" formulation, 11; missile silo technology, 154; nonproliferation programs and, 57–61, 197; Project Sapphire, 98, 165–66; rogue state acquisition concerns, 127–28; Soviet Union dissolution and, 42, 267–68; "Spirit of Pugwash" conference, 10–11; US-Russia Agreements, 43; USSR anthrax and bioweapon concerns, 98–99, 118, 128–29, 165–66; USSR uranium concerns, 97–98. *See also* atomic energy; nonproliferation programs
Weber, Andrew, 97–99, 129–30, 165–68
Weiss, Rai, 141
Western, Karl, 88–89, 189–90, 199–201
Wheeler, John, 77, 212
White, Kendrick, 271
Wiesner, Jerome, 125
Wildermann, Kristin, 64
Wirth, Tim, 157
women, 248–49

Yakobashvili, Zurab, 253
Yatskiv, Yaroslav, 85–86, 207–208, 235–36
Yel'tsin, Boris, 50, 63, 230
Yushchenko, Konstantyn, 235

Zaroubin, Georgi Z., 13
Zel'dovich, Yakov, 76, 77, 201, 205, 240–42
Zelin, Mikhail, 94
Zhdanov, Viktor, 249–50
Zimin, Dmitriy, 236, 239n31
Zimmerman, John, 35–37, 154, 213
zoology, 86–88

GERSON S. SHER is a retired civil servant and foundation executive who has devoted his career to the intersection of scientific cooperation, international affairs, and global security, primarily with the countries of the former Soviet Union.

In the public sector, he was Program Coordinator for US-Soviet and East European Programs at the National Science Foundation, where he served for twenty years and contributed to a variety of other programs, including a presidential initiative with India and policy work in the White House Science Office on scientific communication and national security. He has also worked in several nonprofit organizations, including the National Academy of Sciences, George Soros's International Science Foundation (as Chief Operating Officer) and the US Civilian Research and Development Foundation (now CRDF Global), of which he was Founding President. As President and Executive Vice President of the United States Industry Coalition, he engaged extensively with private US high-tech companies to bring them together with former Soviet weapons of mass destruction scientists to produce mutually profitable civilian technologies. As Senior Advisor at the Henry L. Stimson Center, an independent nonprofit think tank, he worked on projects related to global nuclear security.

Sher received a BA (summa cum laude) in Russian studies from Yale University in 1969 and a PhD in politics from Princeton University in 1975. His doctoral dissertation, *Praxis: Marxist Criticism and Dissent in Socialist Yugoslavia*, was published by Indiana University Press in in 1977. He is the author and translator (from Serbo-Croatian) of numerous books and articles. In June 2008, he received an honorary doctorate from the Moscow Engineering-Physical Institute (MEPhI, now the Russian Federal Nuclear University), Russia's premier university for educating nuclear scientists and engineers. The degree was awarded for the career work he has done to promote science and technology cooperation between the United States and Russia.

www.ingramcontent.com/pod-product-compliance
Lightning Source LLC
Chambersburg PA
CBHW071401300426
44114CB00016B/2143